THE PARTICLE AT THE END OF THE UNIVERSE

THE PARTICLE AT THE END OF THE UNIVERSE

How the Hunt for the Higgs Boson Leads Us to the Edge of a New World

Sean Carroll

DUTTON

DUTTON
Published by Penguin Group (USA) Inc.
375 Hudson Street, New York, New York 10014, U.S.A.
Penguin Group (Canada), 90 Eglinton Avenue East, Suite 700, Toronto, Ontario M4P 2Y3, Canada (a division of Pearson Penguin Canada Inc.); Penguin Books Ltd, 80 Strand, London WC2R 0RL, England; Penguin Ireland, 25 St Stephen's Green, Dublin 2, Ireland (a division of Penguin Books Ltd); Penguin Group (Australia), 707 Collins St., Melbourne, Victoria 3008, Australia (a division of Pearson Australia Group Pty Ltd); Penguin Books India Pvt Ltd, 11 Community Centre, Panchsheel Park, New Delhi – 110 017, India; Penguin Group (NZ), 67 Apollo Drive, Rosedale, Auckland 0632, New Zealand (a division of Pearson New Zealand Ltd); Penguin Books, Rosebank Office Park, 181 Jan Smuts Avenue, Parktown North 2193, South Africa; Penguin China, B7 Jaiming Center, 27 East Third Ring Road North, Chaoyang District, Beijing 100020, China

Penguin Books Ltd, Registered Offices: 80 Strand, London WC2R 0RL, England

Published by Dutton, a member of Penguin Group (USA) Inc.

First printing, November 2012
10 9 8 7 6 5 4 3 2 1

Library of Congress Cataloging-in-Publication Data has been applied for.

ISBN 978-0-525-95359-3

Printed in the United States of America
Set in Adobe Garamond Pro with Rotis
Designed by Daniel Lagin

To Mom,
who took me to the library

People underestimate the impact of a new reality.

—JOE INCANDELA, SPOKESPERSON FOR THE CMS COLLABORATION AT THE LARGE HADRON COLLIDER

CONTENTS

PROLOGUE

JoAnne Hewett is feeling giddy, smiling broadly as she speaks enthusiastically into a video camera. An excited buzz filters up from partygoers at the Swiss consulate in San Francisco. It's a unique event, celebrating the first protons circulating in the underground tunnel of the Large Hadron Collider (LHC) outside Geneva—an enormous particle accelerator on the French-Swiss border that has begun its quest to unlock the secrets of the universe. The champagne flows freely, and no wonder. Hewett's voice rises with emphasis: "I've been waiting for this day for Twenty. Five. Years."

It's a big moment. At this point in 2008, physicists have finally achieved what they have long insisted was necessary to make the next big step forward: a giant accelerator that would smash protons together at very high energies. For a while, they thought the United States was going to build such a machine, but things didn't work out as anticipated. Hewett was just beginning graduate school in 1983, when the U.S. Congress first approved construction of the Superconducting Super Collider (SSC) in Texas. Slated to begin operation before the year 2000, it would have been the largest collider ever built. She, like so many of the brilliant and ambitious physicists of her generation, believed that discoveries there would form the foundation of their research careers.

But the SSC was canceled, pulling the rug out from under physicists who had counted on it to shape the course of their field for decades to come. Politics and bureaucracy and infighting got in the way. Now the LHC, similar in many ways to what the SSC would have been, is at long last about to fire up for the first time, and Hewett and her colleagues are more than ready for it. "What I've done over the past twenty-five years is take every new crazy theory that anybody's ever come up with and calculated its signature [how we identify new particles] at the SSC or LHC," she says.

There is another, more personal reason for Hewett's giddiness. In the video, her red hair is very short, almost a crew cut. It's not a fashion choice. Earlier in the year she was diagnosed with invasive breast cancer, with about a one-in-five chance that it would be terminal. She opted for an extremely aggressive treatment program, involving harsh chemotherapy and a seemingly endless series of surgeries. Her trademark red hair, usually reaching down to her waist, disappeared quickly. At times, she admits with a laugh, she kept her spirits up by thinking about what new particles would be found at the LHC.

JoAnne and I have known each other for years, as friends and colleagues. My own expertise is primarily in cosmology, the study of the universe as a whole, which has recently enjoyed a golden age of new data and surprising discoveries. Particle physics, which has become inseparable from cosmology as an intellectual discipline, has nevertheless been starved for new experimental results that will upend the theoretical applecart and lead us forward to new ideas. The pressure has been building for a long time. Another physicist at the party, Gordon Watts of the University of Washington, was asked whether the long anticipation for the LHC has been stressful. "Oh yeah, completely. I have this shock of gray hair here now. My wife claims it's because of our kid, but it's really because of the LHC."

Particle physics stands on the brink of a new era, in which some theories are going to come crashing down, and perhaps others will turn out to be right on the money. Every physicist at the party has their favorite models—Higgs bosons, supersymmetry, technicolor, extra dimensions, dark matter—a tumble of exotic ideas and fantastic implications.

"My hope for what the LHC will find is 'none of the above,'" Hewett enthuses. "I honestly think it's going to be a surprise, because I think nature is smarter than we are, and she's got some surprises in store for us, and we're going to have a hell of a fun time trying to figure it all out. And it's going to be great!"

That was 2008. In 2012, the San Francisco party to celebrate the inauguration of the LHC is over, and the era of discovery has been officially launched. Hewett's hair has grown back. The treatments were agonizing, but they seem to have worked. And the experiment she's been anticipating for her entire career is making history. After two and a half decades of theorizing, her ideas are finally being tested against real data—particles and interactions never seen before by human beings, surprises that nature has been keeping hidden from us. Until now.

Jump to July 4, 2012, opening day of the International Conference on High Energy Physics. It's a biannual gathering, moving from city to city around the world, this year winding up in Melbourne, Australia. Hundreds of particle physicists, Hewett included, have filled the main auditorium to hear a special seminar. All the investment in the LHC, all the anticipation that has built up over the years, is about to pay off.

The presentation itself is beamed to Melbourne from CERN, the laboratory in Geneva that is home to the LHC. There are two talks, which would ordinarily have been presented in Melbourne as part of the conference program. At the last minute, however, the powers that be decided that a moment of this magnitude should be shared with the many people who had helped make the LHC such a success. The sentiment was appreciated—hundreds of physicists at CERN have lined up for hours before the talks were scheduled to begin at nine a.m., Geneva time, camping out overnight with sleeping bags in hopes of getting a good seat.

Rolf Heuer, director general of CERN, introduces the proceedings. There will be talks by American physicist Joe Incandela and Italian physicist Fabiola Gianotti, the spokespersons for the two major experiments that work to collect and analyze LHC data. Both experiments include more than three thousand collaborators each, most of whom are glued to computer monitors scattered around the globe. The event is being live-streamed, not only to Melbourne, but to anyone who wants to hear the results in real-time. It's an appropriate medium for this celebration of modern Big Science—a high-tech international effort with big stakes and exhilarating rewards.

Traces of nervous energy are evident in both Gianotti's and Incandela's talks, but the presentations speak for themselves. They each give heartfelt thanks to the many engineers and scientists who helped make the experiments possible. Then they carefully explain why we should all believe the results they are about to present, demonstrating that they understand how their machines are working and that the analysis of the data is precise and reliable. Only after the stage has been immaculately set do they show us what they've found.

And there it is. A handful of graphs that wouldn't seem like much to the untutored eye, but with a consistent feature: more events (collections of particles streaming from a single collision) than expected with a certain particular energy. All the physicists in the audience know immediately what it means: a new particle. The LHC has glimpsed a part of nature that had heretofore never been seen. Incandela and Gianotti go through the painstaking statistical analysis meant to separate true discoveries from unfortunate statistical fluctuations, and the results in both cases speak without ambiguity: This is something real.

Applause. In Geneva, Melbourne, and around the world. The data are so precise and clear that even scientists who had worked on the experiments for years are taken aback. Welsh physicist Lyn Evans, who more than anyone else was responsible for guiding the LHC through its rocky path to completion, declared himself "gobsmacked" at the exquisite agreement between the two experiments.

I was at CERN myself that day, masquerading as a journalist in a pressroom next to the main auditorium. Journalists aren't supposed to clap at the news events they cover, but the assembled reporters gave in to the overwhelming emotion of the moment. This wasn't just a success for CERN, or for physics; this was a success for the human race.

We think we know what's been found: an elementary particle called the "Higgs boson," after Scottish physicist Peter Higgs. Higgs himself was in the room for the seminars, eighty-three years old and visibly moved: "I never thought I'd see this happen in my lifetime." Several other senior physicists who had likewise proposed the same idea back in 1964 were also present; the conventions by which theories are named aren't always fair, but this was a moment when everyone could join in the celebration.

So what is the Higgs boson? It's a fundamental particle of nature, of which there aren't many, and a very special kind of particle to boot. Modern particle physics knows of three kinds of particles. There are particles of matter, like electrons and quarks, that constitute the atoms that make up everything we see. There are the force particles that carry gravity and electromagnetism and the nuclear forces, which hold the matter particles together. And then there is the Higgs, in its own unique category.

The Higgs is important not for what it is but for what it does. The Higgs particle arises from a field pervading space, known as the "Higgs field." Everything in the known universe, as it travels through space, moves through the Higgs field; it's always there, lurking invisibly in the background. And it matters: Without the Higgs, electrons and quarks would be massless, just like photons, the particles of light. They would move at the speed of light themselves, and it would be impossible to form atoms and molecules, much less life as we know it. The Higgs field isn't an active player in the dynamics of ordinary matter, but its presence in the background is crucial. Without it, the world would be an utterly different place. And now we've found it.

Some words of caution are in order. What we actually have in hand is evidence for a very Higgs-like particle. It has the right mass, it is produced and decays in roughly the expected ways. But it's too early in the game to say for sure that what we've discovered is definitely the simple Higgs predicted by the original models. It could be something more complicated, or be part of an elaborate web of related particles. But we've definitely found some new particle, and it acts like we think a Higgs boson should. For the purposes of this book, I'm going to treat July 4, 2012, as the day the discovery of the Higgs boson was announced. If reality turns out to be more subtle, then all the better for everyone—physicists live for surprises.

Hopes are high that the Higgs discovery represents the beginning of a new age in particle physics. We know that there is more to physics than we currently understand; studying the Higgs offers a new window into worlds yet unseen. Experimenters like Gianotti and Incandela have a new specimen to study; theorists like Hewett have new clues to build better models. Our understanding of the universe has taken a huge, long-anticipated step forward.

This is the story of the people who have devoted their lives to discovering the ultimate nature of reality, of which the Higgs is a crucial component. There are theorists, sitting with pencil and paper, fueled by espresso and heated disputes with colleagues, turning over abstract ideas in their minds. There are engineers, pushing machines and electronics well beyond the limits of existing technology. And most of all there are the experimenters, bringing the machines and the ideas together to discover something new about nature. Modern physics at the cutting edge involves projects that cost billions of dollars and take decades to complete, requiring extraordinary devotion and a willingness to bet high stakes in search of unique rewards. When it all comes together, the world changes.

Life is good. Have another glass of champagne.

ONE
THE POINT

In which we ask why a group of talented and dedicated people would devote their lives to the pursuit of things too small to be seen.

Particle physics is a curious activity. Thousands of people spend billions of dollars building giant machines miles across, whipping around subatomic particles at close to the speed of light and crashing them together, all to discover and study other subatomic particles that have essentially no impact on the daily lives of anyone who is not a particle physicist.

That's one way of looking at it, anyway. Here's another way: Particle physics is the purest manifestation of human curiosity about the world in which we live. Human beings have always asked questions, and since the ancient Greeks more than two millennia ago, the impulse to explore has grown into a systematic, worldwide effort to discover the basic rules governing how the universe works. Particle physics arises directly from our restless desire to understand our world; it's not the particles that motivate us, it's our human desire to figure out what we don't understand.

The early years of the twenty-first century are a turning point. The last truly surprising experimental result to emerge from a particle accelerator was in the 1970s, more than thirty-five years ago. (The precise date would depend on your definition of "surprising.") It's not because the experimentalists have been asleep at the switch—far from it. The machines have improved by leaps and bounds, reaching into realms that seemed impossibly far away just a short time ago. The problem is that they haven't seen anything we didn't already expect them to see. For scientists, who are always hoping to be surprised, that's extremely annoying.

The problem, in other words, isn't that the experiments have been inadequate—it's that the theory has been too good. In the specialized world of modern science, the roles of "experimentalists" and "theorists" have become quite distinct, especially in particle physics. Gone are the days—as recent as the first half of the twentieth century—when a genius like Italian physicist Enrico Fermi could propose a new theory of the weak interactions, then turn around and guide the construction of the first self-sustained artificial nuclear chain reaction. Today, particle theorists scribble equations on blackboards, which ultimately become specific models, which are tested by experimentalists who gather data from exquisitely precise machines. The best theorists keep close tabs on experiments and vice versa, but no one person is a master of both.

The 1970s saw the finishing touches put on our best theory of particle physics, which goes by the fantastically uninspiring name of the "Standard Model." It's the Standard Model that describes quarks, gluons, neutrinos, and all the other elementary particles you may have heard of. Like Hollywood celebrities or charismatic politicians, scientific theories are put on a pedestal just so we can tear them down. You don't become a famous physicist by showing that someone else's theory is right; you become famous by showing where someone else's theory goes wrong, or by proposing a better theory.

But the Standard Model is stubborn. For decades now, every experiment that we can do here on earth has duly confirmed its predictions. An entire generation of particle physicists has risen up the academic ladder from students to senior professors, all without having a single new phenomenon that they could discover or explain. The anticipation has been close to unbearable.

All this is changing. The Large Hadron Collider represents a new era in physics, smashing together particles with an energy never before achieved by humankind. And it's not just higher energy. It's an energy we've been dreaming about for years, in which we expect to find new theoretically predicted particles and hopefully some surprises—the energy where the force known as the "weak interaction" hides its secrets.

The stakes are high. Peering into the unknown for the first time, anything could happen. There are scads of competing theoretical models hoping to anticipate what the LHC will find. You don't know what you're going to see until you look. At the center of the speculation lies the Higgs boson, an unassuming particle that represents both the last piece of the Standard Model, and the first glimpse into the world beyond.

A big universe made of little pieces

Near the Pacific coast in Southern California, about an hour-and-a-half drive south of where I live in Los Angeles, there is a magical place where dreams come to life: Legoland. At Dino Island, Fun Town, and other attractions, children marvel at an elaborate world constructed from Legos, tiny plastic blocks that can be fitted together in limitless combinations.

Legoland is a lot like the real world. At any moment, your immediate environment typically contains all sorts of substances: wood, plastic, fabric, glass, metal, air, water, living bodies. Very different kinds of things, with very different properties. But when you look more closely, you discover that these substances aren't truly distinct from one another. They are simply different arrangements of a small number of fundamental building blocks. These building blocks are the elementary particles. Like the buildings in Legoland, tables and cars and trees and people represent some of the amazing diversity you can achieve by starting with a small number of simple pieces and fitting them together in a variety of ways. An atom is about one-trillionth the size of a Lego block, but the principles are similar.

We take for granted the idea that matter is made of atoms. It's something we're taught in school, where we do chemistry experiments in classrooms with the periodic table of the elements hanging on the wall. It's easy to lose sight of how amazing that fact is. Some things are hard, some are soft; some things are light, some are heavy; some things are liquid, some are solid, some are gas; some things are transparent, some are opaque; some things are alive, some are not. But beneath the surface, all these things are really the same kind of stuff. There are about one hundred atoms listed in the periodic table, and everything around us is just some combination of those atoms.

The hope that we can understand the world in terms of a few basic ingredients is an old idea. In ancient times, a number of different cultures—Babylonians, Greeks, Hindus, and others—invented a remarkably consistent set of five "elements" out of which everything else was made. The ones we are most familiar with are earth, air, fire, and water, but there was also a heavenly fifth element of aether, or quintessence. (Yes, that's where the movie with Bruce Willis and Milla Jovovich got its name.) Like many ideas, this one was developed into an elaborate system by Aristotle. He suggested that each element sought a particular natural state; for example, earth tends to fall and air tends to rise. By mixing the elements in different combinations, we could account for the different substances we see around us.

Democritus, a Greek philosopher who predated Aristotle, originally suggested that everything we know is made of certain tiny indivisible pieces, which he called "atoms." It's an unfortunate accident of history that this terminology was seized upon by John Dalton, a chemist who worked in the early 1800s, to refer to the pieces that define chemical elements. What we now think of as an atom is not indivisible at all—it consists of a nucleus made of protons and neutrons, around which orbit a collection of electrons. Even the protons and neutrons aren't indivisible; they are made of smaller pieces called "quarks."

The quarks and electrons are the real atoms, in Democritus's sense of indivisible building blocks of matter. Today we call them "elementary particles." Two kinds of quarks—known playfully as "up" and "down"—go into making the protons and neutrons of an atomic nucleus. So, all told, we need only three elementary particles to make up every single piece of matter that we immediately perceive in the environment around us—electrons, up quarks, and down quarks. That's an improvement over the five elements of antiquity, and a big improvement over the periodic table.

Boiling the world down to just three particles is a bit of an exaggeration, however. While electrons and up and down quarks are enough to account for cars and rivers and puppies, they aren't the only particles we've discovered. There are actually twelve different kinds of matter particles: six quarks that interact strongly and get confined inside larger collections like protons and neutrons, and six "leptons" that can travel individually through space. We also have force-carrying particles that hold them together in the different combinations we see. Without force particles, the world would be a boring place indeed—individual particles would just move on straight lines through space, never interacting with one another. It's a fairly small set of ingredients to explain everything we see around us, but frankly, it could be simpler. Modern particle physicists are driven by a desire to do better.

The Higgs boson

That's the Standard Model of particle physics: twelve matter particles, plus a group of force-carrying particles to hold them all together. Not the tidiest picture in the world, but it fits all the data. We have assembled all the pieces needed to successfully describe the world around us, at least here on earth. Out in space we find evidence for things like dark matter and dark energy, stubborn reminders that we certainly don't understand everything yet—these are most certainly *not* explained by the Standard Model.

For the most part the Standard Model divides nicely into matter particles and force-carrying particles. The Higgs boson is different. Named after Peter Higgs, who was one of several people who proposed the idea back in the 1960s, the Higgs boson is somewhat of an ugly duckling. Technically speaking it's a force-carrying particle, but it's a different kind of force carrier from the ones we're most familiar with. From the viewpoint of a theoretical physicist the Higgs seems like an arbitrary and whimsical addition to an otherwise beautiful structure. If it weren't for the Higgs boson, the Standard Model would be the epitome of elegance and virtue; as it is, it's a bit of a mess. And finding the mess-maker has proven to be quite a challenge.

So why were so many physicists convinced that the Higgs boson must exist? You'll hear explanations like "to give mass to other particles" and "to break symmetries," both of which are true but not easy to absorb at first glance. The main point is that without the Higgs boson, the Standard Model would look very different, and not at all like the real world. With the Higgs boson, it's a perfect match.

Theoretical physicists certainly tried their best to come up with theories that didn't have a Higgs boson, or one in which the boson was quite different from the standard story. Many of the theories failed when confronted with the data, and others seemed unnecessarily complicated. None looked like a true upgrade.

And now we've found it. Or something very much like it. Depending on how careful physicists are being, they will say either, "We've discovered the Higgs boson," or "We've discovered a Higgs-like particle," or even "we've discovered a particle that resembles the Higgs." The July 4 announcement described a particle that behaves very much like the Higgs is supposed to behave—it decays into certain other particles in more or less the ways we expect it to. But it's still early, and as we collect more data there is plenty of room for surprises. Physicists don't *want* it to be the Higgs we all expect; it's always more interesting and fun to find something unexpected. There are already tiny hints in the data that this new particle might not be exactly the Higgs we expect. Only further experiments will reveal the truth.

Why we care

I was once interviewed by a local radio station about particle physics, gravitation, cosmology, things like that. It was 2005, the centenary of Albert Einstein's "miraculous year" of 1905, in which he published a handful of papers that turned the world of physics on its head. I did my best to explain some of these abstract concepts, waving my hands up and down, which I can't help but do even when I know I'm on the radio.

The interviewer seemed happy, but after we finished and he was packing up his recording gear, a lightbulb went off in his head. He asked if I would answer one more question. I said sure, and he once again deployed his microphone and headphones. The question was simple: "Why should anybody care?" None of this research is going to lead to a cure for cancer or a cheaper smartphone, after all.

The answer I came up with still makes sense to me: "When you're six years old, *everyone* asks these questions. Why is the sky blue? Why do things fall down? Why are some things hot and others cold? How does it all work?" We don't have to learn how to become interested in science—children are natural scientists. That innate curiosity is beaten out of us by years of schooling and the pressures of real life. We start caring about how to get a job, meet someone special, raise our own kids. We stop asking how the world works, and start asking how we can make it work for us. Later I found studies showing that kids love science up until the ages of ten to fourteen years old.

These days, after pursuing science seriously for more than four hundred years, we actually have quite a few answers to offer the six-year-old inside each of us. We know so much about the physical world that the unanswered questions are to be found in remote places and extreme environments. That's physics, anyway; in fields like biology or neuroscience, we have no difficulty at all asking questions to which the answers are still elusive. But physics—at least the subfield of "elementary" physics, which looks for the basic building blocks of reality—has pushed the boundaries of understanding so far that we need to build giant accelerators and telescopes just to gather new data that won't fit into our current theories.

Over and over again in the history of science, basic research—pursued just for the sake of curiosity, not for any immediate tangible benefit—has proven, almost despite itself, to lead to enormous tangible benefits. Way back in 1831, Michael Faraday, one of the founders of our modern understanding of electromagnetism, was asked by an inquiring politician about the usefulness of this newfangled "electricity" stuff. His apocryphal reply: "I know not, but I wager that one day your government will tax it." (Evidence for this exchange is sketchy, but it's a sufficiently good story that people keep repeating it.) A century later, some of the greatest minds in science were struggling with the new field of quantum mechanics, driven by a few puzzling experimental results that ended up overthrowing the basic foundations of all of physics. It was fairly abstract at the time, but subsequently led to transistors, lasers, superconductivity, light-emitting diodes, and everything we know about nuclear power (and nuclear weapons). Without this basic research, our world today would look like a completely different place.

Even general relativity, Einstein's brilliant theory of space and time, turns out to have down-to-earth applications. If you've ever used a global positioning system (GPS) device to find directions somewhere, you've made use of general relativity. A GPS unit, which you might find in your cell phone or car navigation system, takes signals from a series of orbiting satellites and uses the precise timing of those signals to triangulate its way to a location here on the ground. But according to Einstein, clocks in orbit (and therefore in a weaker gravitational field) tick just a bit faster than those at sea level. A small effect, to be sure, but it builds up. If relativity weren't taken into account, GPS signals would gradually drift away from being useful—over the course of just one day, your location would be off by a few miles.

But technological applications, while important, are ultimately not the point for me or JoAnne Hewett or any of the experimentalists who spend long hours building equipment and sifting through data. They're great when they happen, and we won't turn up our noses if someone uses the Higgs boson to find a cure for aging. But it's not why we are looking for it. We're looking because we are curious. The Higgs is the final piece to a puzzle we've been working on solving for an awful long time. Finding it is its own reward.

The Large Hadron Collider

We wouldn't have found the Higgs without the Large Hadron Collider—another dreary name for an inspiring embodiment of the human passion for discovery. The LHC is the largest, most complex machine ever built by human beings, and it came in at a cool nine billion dollars. The scientists who work at CERN hope it will run productively for fifty years. But they aren't that patient; it would be nice to get some world-changing discoveries right away, thank you very much.

The LHC is gargantuan in every way it can be measured. It was first dreamed up in the 1980s, and approval to start building was given in 1994. Well before it was turned on, the LHC had made big news, as lawsuits attempted to halt its construction on the grounds that it might produce world-consuming black holes. Those were successfully squashed, and the giant collider went to work in earnest in 2009.

Around the world on December 13, 2011, physicists—and quite a few interested onlookers—huddled in seminar rooms and around computer terminals to listen to two talks by researchers from the LHC. The subject was the search for the Higgs boson. This kind of topic is a very frequent subject for physics seminars, and the message is almost always "The search is going well! Wish us luck!" This time was different. Rumors had sped around the Internet for several days before, hinting that we weren't just going to get the usual message—this time, they would be saying, "Okay, we might actually be seeing something. Maybe we've finally found evidence that the Higgs boson is really there."

The answer is yes, there were hints that the LHC was actually seeing the Higgs. Just hints, mind you; not the final word. The LHC smashed protons together at unbelievable energies, and two giant experimental detectors looked at what particles emerged from those collisions; the number of times that two high-energy photons (particles of light) were produced at a certain energy was just a smidgen bigger than we would have expected if there were no Higgs boson. Evidence that something was likely going on, to be sure, but not yet a discovery. But everything smelled right. Rolf Heuer ended the press conference with a flourish: "See you next year with a discovery."

And so they did. On July 4, 2012, two more seminars brought us an update on the search for the Higgs. This time it wasn't a matter of tantalizing hints; they had found a new particle, without question. Thousands of physicists around the world clapped with joy but also exhaled with relief; the LHC was a success.

Crossroads

Particle physics stands at a critical threshold. It's a foundational part of the human race's long-standing quest to better understand how the universe works. It's also very expensive. And its future is unclear.

The search for the Higgs boson isn't just a story of subatomic particles and esoteric ideas. It's also a tale of money, politics, and jealousy. A project that involves so many people, unprecedented international cooperation, and more than a few technological breakthroughs doesn't happen without a certain amount of conniving, dealing, and occasional skullduggery.

The LHC isn't the first giant particle accelerator that aimed to find the Higgs. There was the Tevatron at Fermi National Accelerator Laboratory (Fermilab), just outside Chicago, which turned on in 1983 and finally turned off in September 2011, after a productive lifespan that included the discovery of the top quark—but no Higgs. There was the

Large Electron-Positron collider (LEP), which ran from 1989 to 2000 in the same underground tunnel where the LHC now sits. Rather than colliding relatively massive protons, which tend to create messy splashes of particles when they meet, LEP collided electrons and their antimatter siblings, positrons. That configuration made it possible to do very precise measurements—but none of those measurements revealed the Higgs.

And then there was the Superconducting Super Collider, or SSC, to which Hewett wistfully referred. The SSC was the American version of the LHC—only bigger, better, and scheduled to be ready first. Proposed in the 1980s, the SSC planned to run at energies almost three times as high as the LHC will someday reach (five times as high as it's achieving right now). But the LHC can boast one enormous advantage over the SSC: It got built.

After only a couple of years of running, the LHC has bequeathed to us a genuine discovery, a particle that looks very much like the Higgs boson. It's the end of one era but also the beginning of another. The Higgs is not merely one more particle—it's a special kind of particle, one that can very naturally interact with other kinds of particles we haven't yet detected. We know the Standard Model is not the final answer; the dark matter mapped out by astronomers is clear evidence of that. The Higgs could be the portal that connects our world with another one lurking just out of our reach. Having found a new particle, we have decades of work ahead of us learning about its properties and where else it might lead.

The long-term future of experimental particle physics remains murky. A century or even fifty years ago, it was possible to make a foundational discovery in particle physics with the kind of equipment that could be set up by an individual scientist and a team of students. Those days might be over. If the LHC gives us the Higgs and nothing else, it will be increasingly difficult to convince skeptical governments to allocate even more money to build a next-generation collider.

A machine like the LHC represents an investment of billions of dollars but also of thousands of person-years of effort from dedicated scientists who are devoting their lives to dig just a little bit deeper into nature's mysteries. People like Lyn Evans, who helped build the LHC, or JoAnne Hewett, who studied countless theoretical models, or Fabiola Gianotti and Joe Incandela, who led their experiments to a historic achievement, have placed an enormous wager. They have gambled that this machine will usher in a new age of discovery, and the stakes they've placed are many years of their professional lives. Finding the Higgs is a vindication of all the work they've done. But as Hewett says, what we really want is to be surprised—to discover something nobody anticipated. That's what would really get our minds going.

Historically, nature has been very good at surprising us.

TWO

NEXT TO GODLINESS

In which we explore how the Higgs boson really has nothing to do with God but is nevertheless pretty important.

Leon Lederman has had second thoughts. He knows what he has done, but he can't take it back. It's just one of those small things that has enormous unexpected consequences.

We're speaking, of course, of the "God Particle." Not the particle itself, which is just the Higgs boson. But the *name* "God Particle," for which Lederman is responsible.

One of the world's great experimental physicists, Lederman won the Nobel Prize in Physics in 1988 for discovering that there is more than one type of neutrino. If he hadn't won it for that, he has other achievements that would also be Prize-worthy, including the discovery of a new kind of quark. There are only three known neutrinos and six known quarks, so these kinds of achievements aren't exactly growing on trees. In his spare time he has served as the director of Fermilab and has founded the Illinois Mathematics and Science Academy. Lederman is also a charismatic personality, famous among his colleagues for his humor and storytelling ability. One of his favorite anecdotes relates the time when, as a graduate student, he arranged to bump into Albert Einstein while walking the grounds at the Institute for Advanced Study at Princeton. The great man listened patiently as the eager youngster explained the particle-physics research he was doing at Columbia, and then said with a smile, "That is not interesting."

But in the public eye, Lederman is better known for something less felicitous: saddling the world with the phrase "God Particle" to refer to the Higgs boson. Indeed, that's the title of an engaging book on particle physics and the search for the Higgs that he wrote with Dick Teresi. As the authors explain in the first chapter of their book, they chose the phrase in part because "the publisher wouldn't let us call it the Goddamn Particle, though that might be a more appropriate title, given its villainous nature and the expense it is causing."

Physicists around the world, a notoriously fractious bunch, will happily agree on one thing: They *hate* the name God Particle. Peter Higgs, from whom the more traditional name derives, says with a laugh, "I was really rather annoyed about that book. And I think I'm not the only one."

Meanwhile, journalists around the world, who can be quite contentious in their own right, find unanimity on a single point: They *love* the name God Particle. One of the safest bets in the world is that if you find an article in the popular press about the Higgs boson, at some point the piece will call it the God Particle.

You can hardly blame the journalists. As names go, God Particle is totally box office, while Higgs boson comes off as a bit inscrutable. But you can't blame the physicists, either. The Higgs has nothing whatsoever to do with God. It's just a really important particle, one that's worth getting excited about, even if that excitement doesn't quite rise to the level of religious ecstasy. It's worth understanding why physicists might be tempted to bestow godlike status on this humble elementary particle, even if it's actually free of any theological implications whatsoever. (Does anyone really think God plays favorites among the particles?)

The mind of God

Physicists have a long and complicated relationship with God. Not just with the hypothetical omnipotent being who created the universe, but with the word "God" itself. When they talk about the universe, physicists will often use the idea of God to express something about the physical world. Einstein was famous for this. Among the most frequently repeated quotes from this eminently quotable scientist are "I want to know God's thoughts; the rest are details" and, of course, "I am convinced that God does not play dice with the universe."

Many of us have given into the temptation of following in Einstein's footsteps. In 1992, a NASA satellite called the Cosmic Background Explorer (COBE) released amazing images of tiny ripples in the background radiation left over from the Big Bang. The significance of the event moved George Smoot, one of the investigators behind COBE, to say, "If you're religious, it's like looking at God." And Stephen Hawking, in the concluding paragraph of his mega-selling *A Brief History of Time*, doesn't shy away from using theological language:

> However, if we do discover a complete theory, it should in time be understandable in broad principle by everyone, not just a few scientists. Then we shall all, philosophers, scientists, and just ordinary people, be able to take part in the discussion of the question of why it is that we and the universe exist. If we find the answer to that, it would be the ultimate triumph of human reason—for then we would know the mind of God.

Historically, some of the world's most influential physicists have been quite religious. Isaac Newton, arguably the greatest scientist of all time, was a devout if somewhat heterodox Christian, who spent as much time studying and interpreting the Bible as he did with physics.

In the twentieth century we have the example of Georges Lemaître, a cosmologist who developed the "Primeval Atom" theory—what is now known as the "Big Bang model." Lemaître was a priest as well as a professor at the Catholic University of Leuven, in Belgium. In the Big Bang model, our observable universe began at a singular moment of infinite density about 13.7 billion years ago; in the Christian account, our universe was created by God at some moment in time. There are obvious parallels between the two stories, but Lemaître was always extremely careful not to mix his religion with science. At one point Pope Pius XII tried to suggest that the Primeval Atom could be identified with "Let there be light" from Genesis, but Lemaître himself persuaded him to drop that line of reasoning.

Today, however, most working physicists are much less likely to believe in God than are members of the general public. When you study the workings of the natural world for a living, you tend to be impressed by how well the universe gets along all by itself, without any supernatural assistance. There are certainly prominent examples of religious physicists, but just as certainly the real work of physics gets along without allowing anything other than the natural world into the equation.

God talk

So if physicists aren't big believers in God, why do they keep talking about Him? Two reasons, actually: one good, one less so.

The good reason is simply that God provides a very convenient metaphor for talking about the universe. When Einstein says, "I want to know God's thoughts," he isn't thinking of a literal supernatural being that the pope might be imagining. He's expressing an inner desire to understand the fundamental workings of reality. There is an amazing fact about the universe: It makes sense. We can study what happens to matter under various circumstances, and we find astonishing regularities that never seem to be violated. When these regularities are established as real beyond reasonable doubt, we call them "laws of nature."

The actual laws of nature are very interesting, but it's also interesting that there are laws at all. The laws we've discovered to date take the form of precise and elegant mathematical statements. The physicist Eugene Wigner was so moved by this feature of reality that he spoke of "the unreasonable effectiveness of mathematics in physics." Our universe isn't simply a hodgepodge of stuff doing random things; it's a highly orderly and predictable evolution of fixed constituents of matter, an intricately choreographed dance of particles and forces.

When physicists speak metaphorically of God, they are simply giving in to the natural human tendency to anthropomorphize the physical world—to give it a human face. "God's thoughts" are code for "the underlying laws of nature." We want to know what those laws are. More ambitiously, we'd like to know if those laws could possibly have been different—are the actual laws of nature just one set among many possible ones, or is there something unique and special about our world? We may or may not be able to answer such a grandiose question, but it's the kind of thing that lights the imagination of working scientists.

The other reason scientists succumb to God-talk is a bit less lofty: public relations. Calling the Higgs boson the God Particle might be wildly inaccurate, but it's marketing genius. Physicists react to the God Particle label with horror and disdain. But it draws eyeballs, which is why it will continue to be used, even though every journalist who covers science knows exactly what the physicists think of the term.

"God Particle" gets people to sit up and take notice. Once that phrase has been coined, there's no way it won't be used by everyone trying to explain this esoteric concept to a public with other demands on their attention. Say you are looking for the Higgs boson, and many people will change the channel—maybe the Kardashians have done something outrageous. Say you're looking for the God Particle, and people will at least pay attention when you explain what you mean. The Kardashians will still be acting up tomorrow.

Occasionally this kind of colorful language gets scientists in trouble. In 1993, when the United States was still planning to build a Superconducting Super Collider that would be even more powerful than the LHC, Nobel Laureate Steven Weinberg was testifying before Congress on the virtues of the new machine. At one point the questions took an unexpected turn.

> **Rep. Harris Fawell (R-IL):** I wish sometimes we have some one word that could say it all and that is kind of impossible. I guess perhaps, Dr. Weinberg, you came a little close to it and I'm not sure but I took this down. You said you suspect that it isn't all an accident that there are rules which govern matter and I jotted down, will this make us find God? I'm sure you didn't make that claim, but it certainly will enable us to understand so much more about the universe?
>
> **Rep. Don Ritter (R-PA):** Will the gentleman yield on that? If the gentleman would yield for a moment I would say . . .
>
> **Fawell:** I'm not sure I want to.
>
> **Ritter:** If this machine does that I am going to come round and support it.

Weinberg wasn't so gauche as to refer to the Higgs boson as the God Particle during his Congressional testimony. But the lure of metaphor is strong, and at some point talking about the workings of reality leads one to ask this kind of question.

In case there is any remaining ambiguity: Nothing we might find at the Large Hadron Collider, or might have found at the Superconducting Super Collider, will make us find God. But we will come closer to understanding the ultimate laws of nature.

The final piece

Lederman and Teresi didn't dub the Higgs boson the God Particle just because they knew it would get attention (although the prospect probably crossed their minds). In the end the flamboyant nomenclature attracted as much bad attention as good. As they put it in the preface to a revised edition of their book: "The title ended up offending two groups: 1) those who believe in God, and 2) those who do not. We were warmly received by those in the middle."

What they were trying to do was express the *importance* of the Higgs boson. The book you're reading right now has a slightly more modest title . . . but only slightly. To be honest, physicists don't react with unalloyed approval when I tell them about *The Particle at the End of the Universe*. As far as we know there isn't any "end" to the universe, either at some location in space or at some future moment in time. And if there were a location where the universe could be said to end, there's no reason to think you would find a particle there. And if you did, there's no reason to think it would be the Higgs boson.

But once again, what we're dealing with here is a metaphor. The Higgs isn't located at the spatial or temporal "end of the universe"—it's located at the *explanatory* end. It's the final piece of the puzzle of how the ordinary matter that makes up our everyday world works at a deep level. That's pretty important.

I should quickly rush in with caveats before my fellow physicists get upset again. The Higgs isn't the missing piece of the puzzle of absolutely everything. Finding the Higgs and measuring its properties leaves plenty of physics still to understand. There's gravity, for one thing: an entire force of nature that we can't quite reconcile with the demands of quantum mechanics, and we don't expect the Higgs to be of any help there. There are also dark matter and dark energy, mysterious substances that pervade the universe and yet have resisted direct detection here on earth. There are other hypothetical exotic particles, the kind theoretical physicists love to invent but for which we currently have no evidence. And then, needless to say, there are all the parts of science that present their own challenges without much crucial input from particle physics at all—from atomic and molecular physics, up through chemistry and biology and geology, all the way to sociology and psychology and economics. The human desire to understand the world will not reach a triumphant conclusion just because we have discovered the Higgs boson.

With all those disclaimers out of the way, let's get back to emphasizing the singular role of the Higgs: It's the final part of the Standard Model of particle physics. The Standard Model explains everything we experience in our everyday lives (other than gravity, which is easy enough to tack on). Quarks, neutrinos, and photons; heat, light, and radioactivity; tables, elevators, and airplanes; televisions, computers, and cell phones; bacteria, elephants, and people; asteroids, planets, and stars—all simply applications of the Standard Model in different circumstances. It's the full theory of immediately discernible reality. And it all fits together beautifully, passing a bewildering variety of experimental tests, as long as there is the Higgs boson. Without the Higgs, or something even more bizarre to take its place, the Standard Model wouldn't get off the ground.

Figuring out the trick

There's something fishy about these claims that the Higgs boson is so important. After all, before we actually found it, how did we know it was important at all? What drove us to keep talking about the properties of a hypothetical particle nobody had ever observed?

Imagine you see a performance by a very talented magician, who performs an amazing card trick. The trick involves getting a playing card to mysteriously levitate in the air. You are puzzled by this trick, and you're absolutely sure that the magician didn't actually use mystical powers to make the card levitate. You're also clever and persistent, and after quite a bit of thinking you come up with a way the magician could have done it, involving a thin thread secretly attached to the card. In fact you're able to come up with other possible schemes involving blowing air and heat pumps, but the thread scenario is both simple and plausible. You even go so far as to reproduce the trick at home, convincing yourself that with the right kind of thread you're able to do the trick just like the magician did.

But you go back to catch another performance of the magician's act, where you are able to see the card levitate once again. His version looks just like the one you were able to put together at home—but try as you might, you can't quite see the thread itself.

The Standard Model Higgs boson is like that thread. For a long time we hadn't seen it directly, but we saw its effects. Or even better, we saw features of the world that make perfect sense if it's there, and make no sense without it. Without the Higgs boson, particles such as the electron would have zero mass and move at the speed of light; but instead they do have mass and move more slowly. Without the Higgs boson, many elementary particles would appear identical to one another, but instead they are manifestly different, with a variety of masses and lifetimes. With the Higgs, all these features of particle physics make perfect sense.

In circumstances like these, whether we're thinking about the levitating card or the Higgs boson, there are two options: Our theory is right, or an even more interesting and elaborate theory is right. The effects are there—the card floats, the particles have mass. There must be an explanation. If it's the simple one we'll congratulate ourselves on our cleverness; if it's something more complicated, we'll have learned something very interesting. Maybe the particle we found at the LHC does part of what the Higgs was proposed to do but not all of it; or maybe the job of the Higgs is played by multiple particles, of which we've only found one. We win no matter what, as long as we ultimately succeed in figuring out what's going on.

Fermions and bosons

Let's see if we can't translate this inspirational metaphorical cheerleading about how important the Higgs boson is into a more specific explanation for what it actually is supposed to do.

Particles come in two types: the particles that make up matter, known as "fermions," and the particles that carry forces, known as "bosons." The difference between the two is that fermions take up space, while bosons can pile on top of one another. You can't just take a pile of identical fermions and put them all at the same place; the laws of quantum mechanics won't allow it. That's why collections of fermions make up solid objects like tables and planets: The fermions can't be squeezed on top of one another.

In particular, the *smaller* the mass of the particle, the more space it takes up. Atoms are made out of just three types of fermions—up quarks, down quarks, and electrons—held together by forces. The nucleus, made of protons and neutrons, which in turn are made of up and down quarks, is relatively heavy, and exists in a relatively tiny region of space. The electrons, meanwhile, are much lighter (about 1/2,000th the mass of a proton or neutron) and take up much more space. It's really the electrons in atoms that give matter its solidity.

Bosons don't take up any space at all. Two bosons, or two trillion bosons, can easily sit at exactly the same location, right on top of one another. That's why bosons are force-carrying particles; they can combine to make a macroscopic force field, like the gravitational field that holds us to the earth or the magnetic field that deflects a compass needle.

Physicists tend to use the words "force," "interaction," and "coupling" in practically interchangeable ways. That reflects one of the deep truths uncovered by twentieth-century physics: Forces can be thought of as resulting from the exchange of particles. (As we'll see, that's equivalent to saying "as resulting from vibrations in fields.") When the moon feels the gravitational pull of the earth, we can think of gravitons passing back and forth between the two bodies. When an electron is trapped by an atomic nucleus, it's because photons are exchanged between them. But these forces are also responsible for other particle processes like annihilation and decay, not just pushing and pulling. When a radioactive nucleus decays, we can attribute that event to the strong or weak nuclear force at work, depending on what kind of decay occurs. Forces in particle physics are responsible for a wide variety of goings-on.

Aside from the Higgs, we know four kinds of forces, each with its own associated boson particles. There's gravity, associated with a particle called the "graviton." Admittedly, we haven't actually observed individual gravitons, so the graviton is often not included in discussions of the Standard Model, although we detect the force of gravity every day when we don't all float into space. But given that gravity is a force, the basic rules of quantum mechanics and relativity essentially guarantee that there are associated particles, so we use the word "graviton" to refer to those particles we haven't yet seen on an individual basis. The way that gravity acts as a force on other particles is pretty simple: Every particle attracts every other particle (although very weakly).

Then there is electromagnetism—in the 1800s, physicists figured out that the phenomena of "electricity" and "magnetism" were two different versions of the same underlying force. The particles associated with electromagnetism are called "photons," which we see directly all the time. Particles that do interact via electromagnetism are "charged," while those that don't are "neutral." And just to keep you on your toes, electrical charges can be positive or negative, with like charges pushing each other apart and opposite charges attracting. The ability of like charges to repel each other is absolutely crucial to how the universe works. If electromagnetism were universally attractive, every particle would simply attract every other particle, and all the matter in the universe would do its best to collapse into one giant black hole. Fortunately we have electromagnetic repulsion as well as attraction, which keeps life interesting.

Nuclear forces

Then we have the two "nuclear" forces, so called because (unlike gravity and electromagnetism) they only extend over a very short distance, comparable in size to the nucleus of an atom or less. There is the strong nuclear force, which holds quarks together inside protons and neutrons; its particles are charmingly named "gluons." The strong nuclear force is (unsurprisingly) very strong, and interacts with quarks but not with electrons. Gluons are massless, just like photons and gravitons. When a force is carried by massless particles, we expect its influence to stretch over a very long range, but the strong force is actually very short ranged.

In 1973, David Gross, David Politzer, and Frank Wilczek showed that the strong force has an amazing property: The attraction between two quarks actually grows in strength as the quarks are moved apart. As a consequence, pulling two quarks apart requires more and more energy, so much so that you eventually just create more quarks. It's like pulling on a strip of rubber, with each end representing a quark. You can pull the two ends, but you never get one end all by itself. Instead you create two new ends when the rubber snaps. As a result, you will never see an individual quark alone in the wild; they (and the gluons) are confined inside heavier particles. These composite particles made of quarks and gluons are known as "hadrons," from which the LHC gets its middle name. Gross, Politzer, and Wilczek shared the Nobel Prize in 2004 for this discovery.

Then there is the weak nuclear force, which lives up to its name. Although it doesn't play much of a role in our immediate environment here on earth, the weak force is nevertheless important to the existence of life: It helps the sun shine. Solar energy arises from conversion of protons into helium, which requires turning some of those protons into neutrons, which proceeds by the weak interaction. But down here on earth, unless you're a particle or nuclear physicist, you don't see too much of the weak force in action.

Three different kinds of bosons carry the weak force. There is the Z boson, which is electrically neutral, and there are two different W bosons, one with a positive electric charge and one with a negative electric charge, dubbed W^+ and W^- for short. The W and Z bosons are quite massive by elementary-particle standards (about as heavy as an atom of zirconium, if that's any help), which means that they are hard to produce and decay away fairly quickly, all of which contributes to why the weak interactions are so weak.

In casual speech we use the word "force" to refer to all kinds of things. The force of friction when something is sliding, the force of impact when you smash into a wall, the force of air resistance as a feather falls to the ground. You will have noticed that none of these forces made our list of the four forces of nature, nor do any of them have bosons associated with them. That's the difference between elementary-particle physics and colloquial usage. All of the macroscopic "forces" that we experience as part of our daily routine, from the acceleration when we depress a car's gas pedal to the tug on a leash when a dog suddenly sees a squirrel and takes off, ultimately arise as complicated side effects of the fundamental forces. In fact, with the notable exception of gravity (which is pretty straightforward, pulling everything down), all of those everyday phenomena are just manifestations of electromagnetism and its interactions with atoms. This is the triumph of modern science: to boil the marvelous variety of the world around us down to just a few simple ingredients.

Fields pervade the universe

Of these four forces, one has long stood out as weird: the weak force. Notice that gravity has gravitons, electromagnetism has photons, and the strong force has gluons; one kind of boson for each force. The weak force comes with three different bosons, the neutral Z and the two charged Ws. And these bosons are responsible for strange behaviors, as well. By emitting a W boson one kind of fermion can change into another kind: a down quark can spit out a W^- and change into an up quark. Neutrons, which are made of two downs and an up, decay when they're by themselves outside a nucleus—one of their down quarks emits a W^-, and the neutron converts into a proton, which has two ups and a down. None of the other forces change the identity of the particles they interact with.

The weak interactions, basically, are a mess. And the reason is simple: the Higgs.

The Higgs is fundamentally different from all the other bosons. The others, as we'll see in Chapter Eight, all arise because of some symmetry of nature connecting what happens at different points in space. Once you believe in these symmetries, the bosons are practically inevitable. But the Higgs isn't like that at all. There is no deep principle that requires its existence, but it exists anyway.

After the LHC announced the Higgs discovery on July 4, hundreds of attempts were made at explaining what it was supposed to mean. The biggest reason why this task is such a challenge is that it's not really the Higgs boson itself that is all that interesting; what matters is the Higgs *field* from which the boson arises. It's a fact of physics that all the different particles really arise out of fields—that's quantum field theory, the underlying framework for everything that particle physicists do. But quantum field theory isn't something we teach kids in high school. It's not even something we often discuss in popular physics books; we talk about particles and quantum mechanics and relativity, but we rarely dig into the wonders of quantum field theory underlying it all. When it comes to the Higgs boson, however, it's no longer adequate to skirt around the ultimate field-ness of it all.

When we talk about a "field," we are talking about "something that has some value at every point in space." The temperature of the earth's atmosphere is a field; at every point on the earth's surface (or at any elevation above the surface) the air has a certain temperature. The density and humidity of the atmosphere are likewise fields. But these aren't *fundamental* fields—they are just properties of the air itself. The electromagnetic field or the gravitational field are, in contrast, believed to be fundamental. They're not made of anything else—they are what the world is made of. According to quantum field theory, absolutely everything is made of a field or a combination of fields. What we call "particles" are tiny vibrations in these fields.

This is where the "quantum" part of quantum field theory comes in. There's a lot to say about quantum mechanics, perhaps the most mysterious idea ever to be contemplated by human beings, but all we need is one simple (but hard to accept) fact: How the world appears when we look at it is very different from how it really is.

The physicist John Wheeler once proposed a challenge: How can you best explain quantum mechanics in five words or fewer? In the modern world, it's easy to get suggestions for any short-answer question: Simply ask Twitter, the microblogging service that limits posts to 140 characters. When I posed the question about quantum mechanics, the best answer was given by Aatish Bhatia (@aatishb): "Don't look: waves. Look: particles." That's quantum mechanics in a nutshell.

Every particle we talk about in the Standard Model is, deep down, a vibrating wave in a particular field. The photons that carry electromagnetism are vibrations in the electromagnetic field that stretches through space. Gravitons are vibrations in the gravitational field, gluons are vibrations in the gluon field, and so on. Even the fermions—the matter particles—are vibrations in an underlying field. There is an electron field, an up quark field, and a field for every other kind of particle. Just like sound waves propagate through the air, vibrations propagate through quantum fields, and we observe them as particles.

Just a bit ago we mentioned that particles with a small mass take up more space than ones with a larger mass. That's because the particles aren't really little balls with a uniform density; they're quantum waves. Every wave has a wavelength, which gives us a rough idea of its size. The wavelength also fixes its energy: It requires more energy to have a short wavelength, since the wave needs to change more quickly from one point to another. And mass, as Einstein taught us long ago, is just a form of energy. So lower masses mean less energy mean longer wavelengths mean larger sizes; higher masses mean more energy mean shorter wavelengths mean smaller sizes. It all makes sense once you unpack it.

Stuck away from zero

Fields have a value at every point in space, and when space is completely empty those values are typically zero. By "empty" we mean "as empty as can be," or, more specifically, "with as little energy as it is possible to have." According to that definition, fields like the gravitational field or the electromagnetic field sit quietly at zero when space is truly empty. When they're at some other value, they carry energy, and therefore space isn't empty. All fields have tiny vibrations because of the intrinsic fuzziness of quantum mechanics, but those are vibrations around some average value, which is typically zero.

The Higgs is different. It's a field, just like the others, and it can be zero or some other value. But it doesn't *want* to be zero; it wants to sit at some constant number everywhere in the universe. The Higgs field has less energy when it's nonzero than when it's zero.

As a result, empty space is full of the Higgs field. Not a complicated set of vibrations that would represent a collection of individual Higgs bosons; just a constant field, sitting quietly in the background. It's that ever-present field at every point in the universe that makes the weak interactions what they are and gives masses to elementary fermions. The Higgs boson—the particle discovered at the LHC—is a vibration in that field around its average value.

Because the Higgs particle is a boson, it gives rise to a force of nature. Two massive particles can pass by each other and interact by exchanging Higgs bosons, just like two charged particles can interact by exchanging photons. But this Higgs force is *not* what gives particles mass, and it's generally not what all the fuss is about. What gives

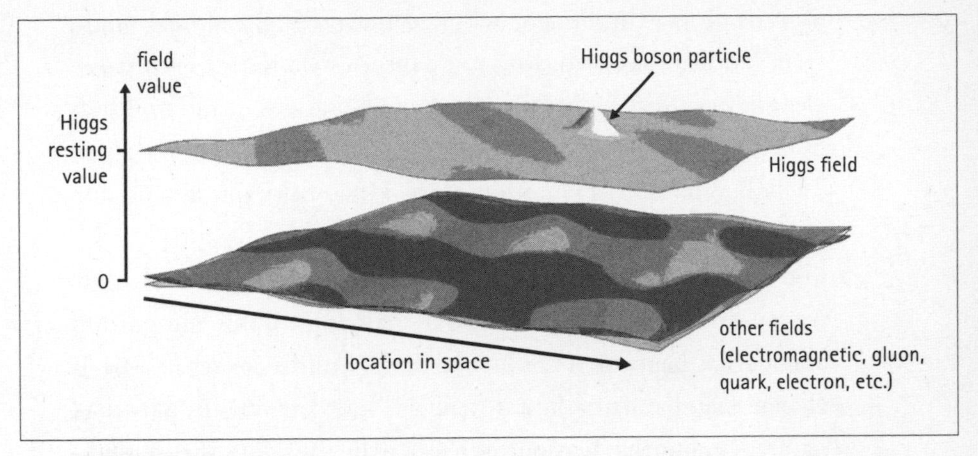

One major difference between the Higgs field and other fields is that the resting value of the Higgs is away from zero. All fields undergo tiny vibrations due to the intrinsic uncertainties of quantum mechanics. A larger vibration appears to us as a particle, in this case the Higgs boson.

particles mass is this Higgs field sitting quietly in the background, providing a medium through which other particles move, affecting their properties along the way.

As we travel through space, we're surrounded by the Higgs field and moving within it. Like the proverbial fish in water, we don't usually notice it, but that field is what brings all the weirdness to the Standard Model.

Executive summary

There is a great deal of profound and challenging physics associated with the idea of the Higgs boson. But for right now let's just give the overall summary of how the Higgs field works and why it's important. Without further ado:

- The world is made of *fields*—substances spread through all of space that we notice through their vibrations, which appear to us as particles. The electric field and the gravitational field might seem familiar, but according to quantum field theory even particles like electrons and quarks are really vibrations in certain kinds of fields.

- The Higgs boson is a vibration in the Higgs field, just as a photon of light is a vibration in the electromagnetic field.
- The four famous forces of nature arise from *symmetries*—changes we can make to a situation without changing anything important about what happens. (Yes, it makes no immediate sense that "a change that doesn't make a difference" leads directly to "a force of nature" . . . but that was one of the startling insights of twentieth-century physics.)
- Symmetries are sometimes *hidden* and therefore invisible to us. Physicists often say that hidden symmetries are "broken," but they're still there in the underlying laws of physics—they're simply disguised in the immediately observable world.
- The weak nuclear force, in particular, is based on a certain kind of symmetry. If that symmetry were unbroken, it would be impossible for elementary particles to have *mass*. They would all zip around at the speed of light.
- But most elementary particles do have mass, and they don't zip around at the speed of light. Therefore, the symmetry of the weak interactions must be broken.
- When space is completely empty, most fields are turned off, set to zero. If a field is not zero in empty space, it can break a symmetry. In the case of the weak interactions, that's the job of the Higgs field. Without it, the universe would be an utterly different place.

Got all that? It's a bit much to swallow, admittedly. It will make more sense when we complete our journey through the rest of the chapters. Trust me.

The rest of the book will be a back-and-forth journey through the ideas behind the Higgs mechanism and the experimental quest to discover the boson. We'll start with a quick overview of how the particles and forces of the Standard Model fit together, then explore the astonishing ways in which physicists use technology and gumption to discover new particles. After that it's back to theory, as we think about fields and symmetries and how the Higgs can hide symmetries from our view. Finally we can show how the Higgs was discovered, how the news was spread, who will get the credit, and what it means for the future.

It should be clear why Leon Lederman thought that the God Particle was an appropriate name for the Higgs boson. That boson is the hidden piece of equipment that explains the magic trick the universe is pulling on us, giving particles different masses and thereby making particle physics interesting. Without the Higgs, the intricate variety of the Standard Model would collapse to a featureless collection of pretty much identical particles, and all of the fermions would be essentially massless. There would be no atoms, no chemistry, no life as we know it. The Higgs boson, in a very real sense, is what brings the universe to life. If there were one particle that deserved such a lofty title, there's no question it would be the Higgs.

THREE

ATOMS AND PARTICLES

In which we tear apart matter to reveal its ultimate constituents, the quarks and the leptons.

In the early 1800s, German physician Samuel Hahnemann founded the practice of homeopathy. Dismayed by the ineffectiveness of the medicine of his time, Hahnemann developed a new approach based on the principle of "like cures like"—a disease can be treated by precisely the same substance that causes it in the first place, as long as that substance is properly manipulated. The way to manipulate it is known as "potentization," which consists of diluting the substance repeatedly in water, shaking vigorously each time. A typical method of dilution might mix one part of substance and ninety-nine parts water. You prepare a homeopathic remedy by diluting, shaking, diluting again, shaking again, as many as two hundred times.

More recently, Crispian Jago, a professional software consultant and recreational skeptic, wanted to demonstrate that he doesn't believe homeopathy is a valid approach to medicine. So he decided to apply the method of serial dilution to an easily obtained substance: his own urine. Which he then proceeded to drink. Because he was a bit impatient, he only diluted the urine thirty times. Except that he didn't call it "urine," he called it "piss," and then proclaimed that he was developing a cure for being pissed, which translates either as angry (for those in the U.S.)

or inebriated (for those in the U.K.). The results, naturally enough, were presented to the world in the form of a boisterous YouTube video.

Jago had good reason for being undisturbed by the prospect of drinking urine that had been diluted in a 1:99 concentration thirty times over: By the time he got to the final glass, there was none of the original stuff left. Not just "a minuscule amount" but really none at all, if his dilutions were sufficiently careful.

That's because everything in our everyday world—urine, diamonds, french fries, really everything—is made of atoms, usually combined into molecules. Those molecules are the smallest unit of a substance that can still be thought of as that substance. Separately, two hydrogen atoms and one oxygen atom are just atoms; combined, they become water.

Because the world is made of atoms and molecules, you can't dilute things forever and have them maintain their identity. A teaspoonful of urine might contain approximately 10^{24} molecules. If we dilute once by mixing one part urine with ninety-nine parts water, we're left with 10^{22} urine molecules. Dilute twice and we have 10^{20} molecules. By the time we've diluted twelve times, on average there's only one molecule of the original substance remaining. After that, it's all window dressing—we're just mixing water into more water. With about forty dilutions we could dilute away every molecule in the known universe.

So when Jago finished the procedure and took his final triumphant swig, the water he was drinking was as pure as any that would ordinarily come out of the tap. Advocates of homeopathy know this, of course. They believe that the water molecules retain a "memory" of whatever herb or chemical was used in the original dilution, and indeed that the final solution is more potent than the substance was to start. This violates everything we know about physics and chemistry, and clinical trials rate homeopathic remedies no better than placebos at combating disease. But everyone is entitled to their own opinion.

We are not, as the saying goes, entitled to our own facts. And the fact that matter is made of atoms and molecules is a striking one. Really there are two critical facts: first, that we can take matter and break it up into little chunks that represent the smallest possible unit of that kind of thing; and second, that it only takes a few fundamental building blocks combined in different ways to account for all the variety of the observable world.

At first glance the particle zoo can seem complex and intimidating, but there are only twelve matter particles, which fall neatly into two groups of six: quarks, which feel the strong nuclear force, and leptons, which do not. It's an amazing story, put together over the course of a century, from the discovery of the electron in 1897 to the detection of the last elementary fermion (the tau neutrino) in 2000. Here we'll take a whirlwind tour, saving the quantitative details for Appendix Two. When the smoke clears we will have a relatively manageable collection of particles from which everything else is made.

Pictures of atoms

Everyone has seen cartoon images of atoms. They are usually portrayed as tiny solar systems, with a central nucleus surrounded by orbiting electrons. It's an iconic image, which serves, for example, as the logo of the U.S. Atomic Energy Commission. It's also misleading in a subtle way.

The cartoon atom represents the Bohr model, named after Danish physicist Niels Bohr, who applied insights from the early days of quantum mechanics to the model of atoms that had been previously developed by New Zealand–born British physicist Ernest Rutherford. In Rutherford's version of the atom, electrons orbit the nucleus at any distance you might imagine, just like planets in the real solar system (except they are attracted to the center by electromagnetism, not by gravity). Bohr modified that idea by insisting that the electrons can travel only on certain particular orbits, which was a great step forward in fitting the data from radiation emitted by atoms. These days we

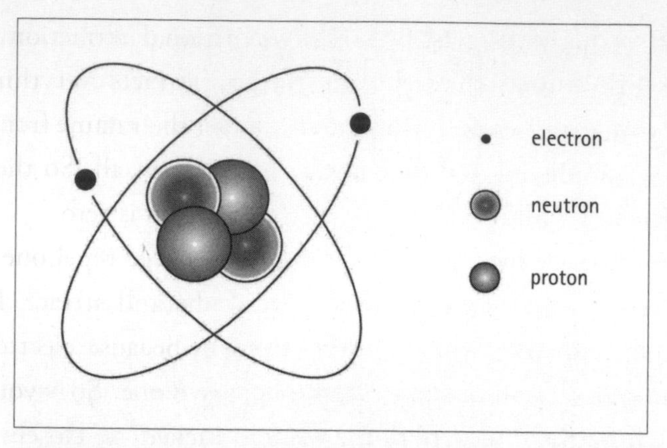

Cartoon image of an atom; in this case, helium. A nucleus consisting of two protons and two neutrons sit at the center, while two electrons "orbit" on the outskirts.

know that the electrons don't really "orbit" at all, because they don't really have a "position" or "velocity." Quantum mechanics says that the electrons persist in clouds of probability known as "wave functions," which tell us where we might find the particle if we were to look for it.

All that being granted, the basic cartoon we have in mind of what an atom looks like isn't that bad, if what we're looking for is some intuitive grasp of what's going on. Nucleus in the middle, electrons on the outskirts. The electrons are relatively light; more than 99.9 percent of the mass of an atom is located in the nucleus. That nucleus is made of a combination of protons and neutrons. A neutron is a bit heavier than a proton—a neutron is about 1,842 times as heavy as an electron, while a proton is about 1,836 times as heavy. Protons and neutrons are both called "nucleons," as they are the particles that make up nuclei (plural of "nucleus"). Aside from the fact that the proton has an electric charge and the neutron is a bit heavier, the two nucleons are remarkably similar particles.

Like many things in life, the nature of an atom is one of exquisite balance. The electrons are attracted to the nucleus by the force of electromagnetism, which is enormously stronger than the force of gravity. The electromagnetic attraction between an electron and a proton is about 10^{39} times stronger than the gravitational attraction between them. But while gravity is simple—everything attracts everything else—electromagnetism is more subtle. Neutrons get their name from the fact that they are neutral, having no electric charge at all. So the electromagnetic force between an electron and a neutron is zero.

Particles with the same kind of electric charge repel one another, while opposites live up to the romantic cliché and attract. Electrons are attracted to the protons inside a nucleus because electrons carry a negative charge and protons carry a positive one. So—you may be asking yourself—why don't the protons packed so closely inside a nucleus push one another apart? The answer is that their mutual electromagnetic repulsion does indeed push them apart, but it is overwhelmed by the strong nuclear force. Electrons don't feel the strong force (just like neutrons don't feel electromagnetism), but protons and neutrons do, which is why they can combine to make atomic nuclei. Only up to a point, however. If the nucleus gets too big, the electric repulsion just becomes too much to take, and the nucleus becomes radioactive; it may survive for a while, but eventually it will decay into smaller nuclei.

Antimatter

Everything you see around you right now, and everything you have ever seen with your own eyes, and everything you have ever heard with your ears and experienced with any of your senses, is some combination of electrons, protons, and neutrons, along with the three forces of gravity, electromagnetism, and the nuclear force that holds protons and neutrons together. The story of electrons, protons, and neutrons had come together by the early 1930s. At that time, it must have been irresistible to imagine that these three fermions were really the fundamental ingredients of the universe, the basic Lego blocks out of which everything is constructed. But nature had some more twists in store.

The first person to understand the basic way fermions work was British physicist Paul Dirac, who in the late 1920s wrote down an equation describing the electron. An immediate consequence of the Dirac equation, although one that took physicists a long time to accept, is that every fermion is associated with an opposite type of particle, called its "antiparticle." The antimatter particles have exactly the same mass as their matter counterparts, but an opposite electric charge. When a particle and an antiparticle come together, they typically annihilate into energetic radiation. A collection of antimatter is therefore a great way (in theory) to store energy, and has fueled much speculation about advanced rocket propulsion in science-fiction stories.

Dirac's theory became a reality in 1932, when American physicist Carl Anderson discovered the positron, the antiparticle of the electron. There is a tight symmetry between matter and antimatter; a person made of antimatter would undoubtedly call the particles of which they were made "matter," and accuse us of being made of antimatter. Nevertheless, the universe we observe is full of matter and contains very little antimatter. Exactly why that should be so remains a mystery to physicists, although we have a number of promising ideas.

Anderson was studying cosmic rays, high-energy particles from space that crash into the earth's atmosphere, producing other particles that eventually reach us on the ground. It's like you're using the air above you as a giant particle detector.

To create images of the tracks of charged particles, Anderson used an amazing technology known as the "cloud chamber." It's an apt name, as the basic principle is similar to that of the actual clouds we see in the sky. You fill a chamber with gas that is supersaturated with water vapor. "Supersaturated" means that the water vapor really wants to form into droplets of liquid water, but it won't do it without some external nudge. In a regular cloud, the nudge typically comes in the form of some speck of impurity, such as dust or salt. In a physicist's cloud chamber, the nudge comes when a charged particle passes through. The particle bumps into the atoms inside the chamber, shaking loose electrons and creating ions. Those ions serve as nucleation sites for tiny droplets of water. So a passing charged particle will leave a trail of droplets in its wake, much like the contrail created by an airplane, lingering evidence of its passage.

Anderson took his cloud chamber, wrapped in a powerful magnet, up to the roof of the aeronautics building at the California Institute of Technology, or Caltech, and watched for cosmic rays. Obtaining the properly supersaturated vapor inside required a rapid decrease in pressure, caused by a piston that would cause a loud bang each time it was released. The chamber was only operated at night due to its massive electricity consumption. Bangs would reverberate through the Pasadena air every evening, noisy testimony that secrets of the universe were being discovered.

The pictures Anderson took showed an equal number of particles curving clockwise and counterclockwise. The obvious explanation was that there were just as many protons as electrons contained in the radiation; indeed, you might expect exactly that, since negatively charged particles can't be created without also creating a balancing positive charge. But Anderson had another piece of data he could use: the thickness of the ion trail left in his cloud chamber. He recognized that, given the curvature of the tracks, any protons that would produce them would have to be relatively slow-moving. (In this context, that means "slower than 95 percent the speed of light.") In that case, they would leave thicker ion trails than what was observed. It seemed that the mysterious particles passing through the chamber were positively charged, like a proton, but relatively light, like an electron.

There was one other logical possibility: Maybe the tracks were simply electrons moving backward. To test this idea, Anderson introduced a plate of lead bisecting the chamber. A particle moving from one side of the lead to the other would slow down just a bit, clearly indicating the direction of its trajectory. In an iconic image from the history of particle physics, we see a counterclockwise-curving particle moving through the cloud chamber, passing through the lead, and slowing

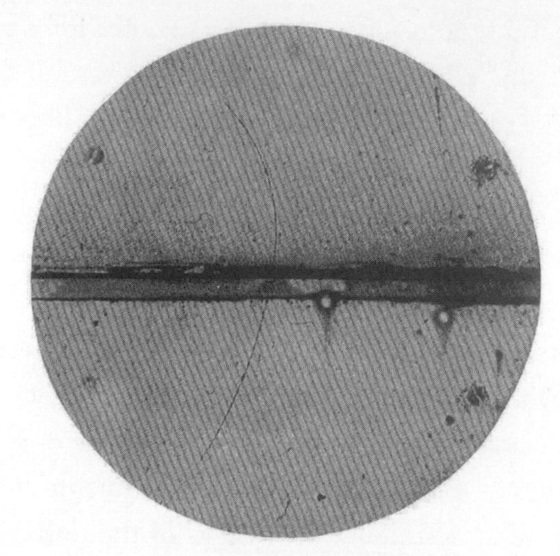

The cloud-chamber image from the discovery of the positron by Carl Anderson. The path of the positron is the curved line that starts near the bottom, hits the lead plate in the middle, and curves more sharply as it continues toward the top.

down afterward—the discovery of the positron. Giants of the field, such as Ernest Rutherford, Wolfgang Pauli, and Niels Bohr, were incredulous at first, but a beautiful experiment will always win out over theoretical intuition, no matter how brilliant. The idea of antimatter had entered the world of particle physics for good.

Neutrinos

So instead of just three fermions (proton, neutron, electron), we have three more (antiproton, antineutron, positron) for a total of six—still fairly parsimonious. But nagging problems remained. For example, when neutrons decay, they turn into protons by emitting electrons. Careful measurements of this process seemed to indicate that energy was not conserved—the total energy of the proton and electron was always a bit less than that of the neutron from which they came.

The answer to this puzzle was suggested in 1930 by Wolfgang

Pauli, who realized that the extra energy could be carried off by a tiny neutral particle that was hard to detect. He called his idea the "neutron," but that was before the name was attached to the heavy neutral particle we find in nuclei. After that happened, to stave off confusion Enrico Fermi dubbed Pauli's particle the "neutrino," from the Italian for "little neutral one."

In fact the decay of a neutron emits what we now recognize as an antineutrino, but the principle was absolutely right. Pauli was quite embarrassed at the time for suggesting a particle that didn't seem detectable, but these days neutrinos are bread and butter for particle physicists (as is proposing hard-to-observe hypothetical particles).

There was still the question of the exact process by which neutrons decay. When particles interact with one another, that implies some kind of force, but the decay of a neutron wasn't what we would expect from gravity, electromagnetism, or the nuclear force. So physicists started attributing neutron decay to the "weak nuclear force," because it obviously had something to do with nucleons but also obviously wasn't the force holding nuclei together, which was dubbed the "strong nuclear force."

The existence of the neutrino established a nice little symmetry among the elementary particles. There were two light particles, the

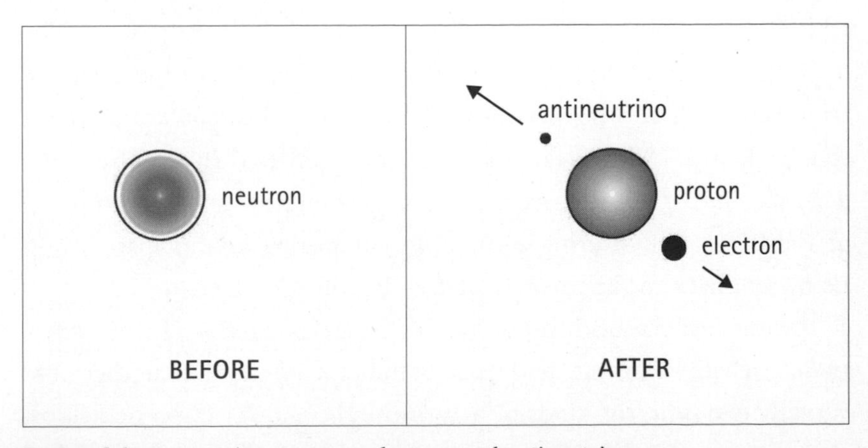

Decay of the neutron into a proton, electron, and antineutrino.

electron and neutrino, which were eventually dubbed "leptons" from the ancient Greek word for "small." And there were two heavy particles, the proton and neutron, which were (somewhat later on) dubbed "hadrons" from the ancient Greek for "large." The hadrons feel the strong nuclear force, while the leptons do not. Each category contained one charged particle and one neutral one. You could be forgiven for thinking that we had it nailed down.

Generations

Then in 1936, a visitor dropped in from the sky—the muon. Carl Anderson, discoverer of the positron, and Seth Neddermeyer were again studying cosmic rays. They found a particle that is negatively charged like the electron but heavier, although lighter than an antiproton would be. It was dubbed the "mu meson," but physicists later realized that it wasn't a meson (which is a boson made of a quark and an antiquark) at all, so the name was shorted to "muon." For a time in the 1930s, fully half of the known elementary particles (electron, positron, proton, neutron, muon, and antimuon) had been discovered in Carl Anderson's lab at Caltech. Who knows? Maybe a decade or two from now, half of the by-then-known elementary particles will have been discovered at the LHC.

The muon was a complete surprise. We already had the electron; why should it have a heavier cousin? Physicists' bafflement was captured succinctly in I. I. Rabi's famous quip, "Who ordered that?" This is exactly the kind of response we're eventually hoping for from experiments at the LHC—discovering something completely unanticipated, and being sent back to the theoretical drawing board as a result.

It was just the beginning. In 1962, experimentalists Leon Lederman, Melvin Schwartz, and Jack Steinberger showed that there are actually two different kinds of neutrinos. There are electron neutrinos, which interact with electrons and are often created along with them, but also muon neutrinos, which go hand in hand with muons. When the neutron decays, it emits an electron, a proton, and an electron antineutrino; when the muon itself decays, it emits an electron and an electron antineutrino, but also a muon neutrino.

And then the process repeated. In the 1970s, the tau particle was discovered, also negatively charged like the electron but even heavier than the muon. These three particles turn out to be almost identical cousins, differing only in mass. In particular, all of them feel the weak and electromagnetic forces but not the strong interaction. And the tau has its own kind of neutrino, which was long anticipated but not directly detected until the year 2000.

We've worked our way up to no fewer than six leptons, which come in three "families" or "generations": the electron and its neutrino, the muon and its neutrino, and the tau and its neutrino. It's perfectly natural to wonder whether there is a fourth generation or beyond lurking out there. Right now the answer is a definite maybe, although there is evidence that three generations are all we get. That's because the known

	FIRST GENERATION	SECOND GENERATION	THIRD GENERATION
charged leptons (charge -1)	electron	muon	tau
neutrinos (charge 0)	electron neutrino	muon neutrino	tau neutrino

The leptons of the Standard Model, arranged into three generations. Larger circles indicate more massive particles, although the sizes are not to scale.

neutrinos have very small masses—certainly much lighter than the electron. We now know how to search for new light particles, by carefully analyzing the decays of heavier ones. We can count how many neutrino-like particles there must be to account for those decays, and the answer is three. It's impossible to be sure that there aren't more lurking out there, perhaps with anomalously large masses, but it may be that we've found all the neutrinos (and therefore all the generations of leptons) there are to find.

Quarks and hadrons

Meanwhile the hadrons were not exactly sitting still. The mid-century advent of particle accelerators led to a boom in the number of so-called elementary particles that physicists had discovered. There were pions, kaons, eta mesons, rho mesons, hyperons, and more. Willis Lamb, during his own Nobel lecture in 1955, cracked, "The finder of a new elementary particle used to be rewarded by a Nobel Prize, but such a discovery now ought to be punished by a ten-thousand-dollar fine."

All of these new particles were hadrons—unlike the leptons, they interacted strongly with neutrons and protons. Increasingly physicists began to suspect that the newcomers weren't really "elementary" at all, but rather reflected some deeper underlying structure.

The code was finally cracked in 1964 by Murray Gell-Mann and George Zweig, who independently proposed that hadrons were made of smaller particles called "quarks." Just like leptons, they come in six different flavors: up, down, charm, strange, top, and bottom. The up/charm/top quarks all have electric charge +2/3, while the down/strange/bottom quarks have charge −1/3; these are sometimes grouped as "up-type" and "down-type" quarks, respectively.

Unlike leptons, each flavor of quark really represents a triplet of particles, rather than just one. The three kinds of each quark are fancifully labeled after colors: red, green, or blue. The names are fun, not

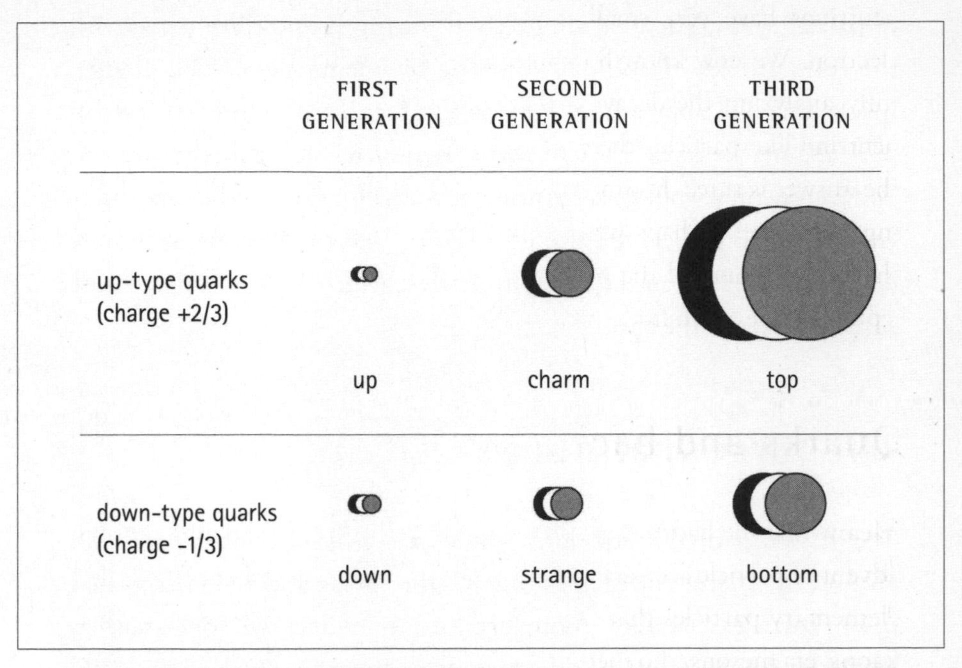

The quarks of the Standard Model, arranged into three generations. Each type of quark comes in three colors. Larger circles indicate more massive particles, although the sizes are not to scale.

realistic; you can't actually see quarks, and if you could they wouldn't actually have those colors.

Quarks are "confined," which means that they exist only in combinations inside hadrons, never isolated by themselves. When they combine, it is always into "colorless" combinations. Protons and neutrons each have three quarks inside: A proton is two ups and a down, while a neutron is two downs and an up. One of those quarks will be red, one will be green, and one will be blue; together they make white, which counts as colorless by the terms of this analogy. Later we will see that there are also "virtual" quark-antiquark pairs popping in and out of existence inside the nucleons, but they come in color-anticolor combinations, leaving the overall whiteness unaffected.

It's impossible to look at the figures portraying the leptons and quarks without noticing some patterns. In both cases we have six types of particles. And these six are precisely arranged into three pairs, with the two particles in each pair differing by one unit of electric charge. Might there be some deeper explanation for this structure? At least in part, the answer is yes. The two particles in each pair, such as an electron and its neutrino, would be identical if it wasn't for the meddlesome influence of the Higgs field filling empty space. That's a reflection of the role of the Higgs as a breaker of symmetries, which we'll examine more carefully later in the book.

The force that doesn't fit

The fermions of the Standard Model are what give the matter all around us its size and shape. But it's the forces and their associated boson particles that allow those fermions to interact with one another. Fermions can push or pull on one another by tossing bosons back and forth, or they can lose energy or decay into other fermions by spitting out some kind of boson. Without the bosons, the fermions would simply move along straight lines for all eternity, unaffected by anything else in the universe. And the reason why the universe is so bloody complex and interesting is that these forces are all different, and push and pull in complementary ways.

Physicists often say that there are four forces of nature—they don't include the Higgs, and not just because it took a long time to discover it. The Higgs is different from the other bosons. The others are what are called "gauge bosons"—as we'll discuss in Chapter Eight, they are deeply related to underlying symmetries of nature. The graviton is a bit different from the others. Every elementary particle has a certain intrinsic "spin," and the photon, gluons, and W/Z bosons all have a spin equal to one, while the graviton has a spin of two. (See Appendix One for some details.) We don't yet know how to reconcile gravity with the demands of quantum mechanics, but it's still fair to call it a "gauge boson."

The Higgs, on the other hand, is completely different. It's what we call

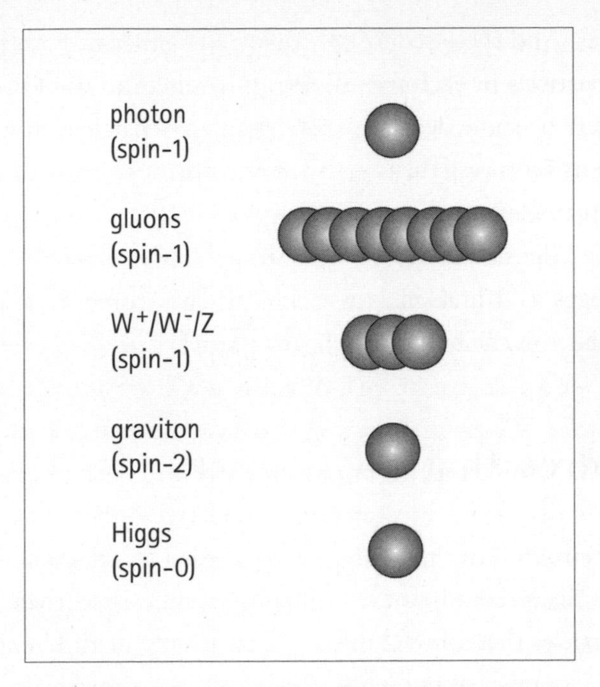

The bosons of the Standard Model. (In this book we include gravitons, although not everyone does.) All the bosons are electrically neutral except for the Ws, and all are massless except for the Ws, Z, and the Higgs.

a "scalar" boson, which means it has zero spin. Unlike the gauge bosons, the Higgs is not forced on us by a symmetry or any other deep principle of nature. A world without the Higgs would look very different, but it would be perfectly consistent as a physical theory. As important as it is, the Higgs is somewhat of a blemish on the beautiful mathematical structure of the Standard Model. Nevertheless, it is a boson, and therefore it can be exchanged back and forth by other particles, giving rise to a force of nature.

The Higgs boson is a vibration in the Higgs field, and the Higgs field is what gives mass to all of the massive elementary particles. So the Higgs boson interacts with all of the massive particles in our zoo—the quarks, the charged leptons, and the W and Z bosons. (Neutrino masses aren't completely understood as yet, so let's pretend that they don't interact with the Higgs, although the jury is still out.) And the more massive a particle is, the more strongly it couples to the Higgs. Really it's the other way around: The more strongly a particle couples to the Higgs, the more mass it picks up by moving through the ambient Higgs field that pervades empty space.

This feature of the Higgs—it interacts more strongly with more massive particles—is absolutely crucial when it comes to studying the beast at the LHC. The Higgs is a heavy particle itself, and even when we produce it we aren't able to see it directly; it will very rapidly decay into other particles. We expect there will be a certain rate of decay into (for example) W bosons, a different rate of decay into bottom quarks, a different rate of decay into tau mesons, and so forth. And it's not random—we know exactly how the Higgs is supposed to interact with other particles (because we know how much mass they each have), so we can calculate quite precisely the expected frequency of different kinds of decays.

What we really want is to be wrong. It's a great triumph to discover the Higgs, but things get really exciting when we are surprised by something new. Searching for invisible particles that are hard to produce and decay quickly into other particles is a challenging task. It's a matter of patience, precision, and careful statistical analysis. The good news is that the laws of physics (or any one hypothetical version of them) are unforgiving; the predictions for what we should see are unambiguous and inalterable. If the Higgs turns out to be different from what we expect, it will be a clear sign that the Standard Model has finally failed us, and the door to new phenomena has been opened at last.

FOUR
THE ACCELERATOR STORY

In which we trace the colorful history of the unlikely pastime of smashing together particles at ever-higher energies.

When I was about ten years old, I discovered the science section in our local library in Lower Bucks County, Pennsylvania. I was immediately hooked. My favorite books were in astronomy and physics—the 520s and 530s, according to the venerable Dewey Decimal System. One of the books I pored over most intently was a modest volume entitled *High Energy Physics*, by Hal Hellman. I was doing my reading in the late 1970s, but the book had been written in 1968, before the Standard Model was put together—back when quarks were exotic and somewhat scary-sounding theoretical speculations. But hadrons had been discovered in abundance, and *High Energy Physics* was full of evocative photographs of particle tracks, each representing a fleeting glimpse of nature's secrets.

Many of these photographs had been taken at the mighty Bevatron, one of the leading particle accelerators of the 1950s and '60s. The Bevatron was located in Berkeley, California, but that's not where the name came from; it was derived from "billions of electron volts," the energy the accelerator was able to reach. (As we'll explain below, an electron volt is a weird unit of energy much beloved by particle physicists.) One billion corresponds to the prefix "giga–," so a billion electron volts is one GeV, but back in those days Americans would often use "BeV," and besides, "Gevatron" just doesn't sound right.

The Bevatron contributed to two Nobel Prizes: in 1959 to Emilio Segrè and Owen Chamberlain, for the discovery of the antiproton, and in 1968 to Luis Alvarez, for the discovery of too many particles to count—all those pesky hadrons. Sometime later, Alvarez and his son Walter were the ones who first demonstrated that an asteroid impact was the likely cause of the extinction of dinosaurs, by discovering an anomalously high concentration of iridium in geological strata that formed around that time.

The idea behind particle accelerators is simple: Take some particles, accelerate them to very high velocities, and slam them into some other particles, watching carefully to see what comes out. The procedure has been compared to smashing together two fine Swiss watches and trying to figure out what they are made of by watching the pieces fly apart. Unfortunately, this analogy has it backward. When we smash particles together, we're not looking for what they are made of; we're trying to create brand-new particles that weren't there before we did the smashing. It's like smashing together two Timex watches and hoping that the pieces assemble themselves into a Rolex.

To attain these velocities, accelerators use a basic principle: Charged particles (such as electrons and protons) can be pushed around by electric and magnetic fields. In practice, we use electric fields to accelerate particles to ever-higher speeds, and magnetic fields to keep them moving in the right direction, such as around the circular tubes of the Bevatron or the LHC. By delicately tuning these fields to push and nudge particles in just the right way, physicists can reproduce conditions that would otherwise never be seen here on earth. (Cosmic rays from outer space can be even more energetic, but they are also rare and hard to observe.)

The technological challenge is clear: Accelerate particles to as high an energy as we can, smash them together, and look to see what new particles are created. None of these steps is easy. The LHC represents

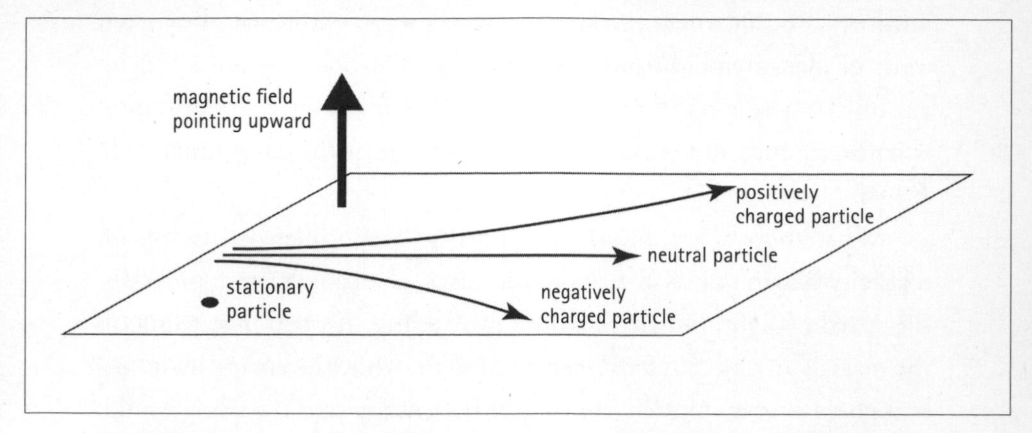

The influence of a magnetic field on moving particles. If the magnetic field is pointing upward, it pushes positively charged particles in a counterclockwise direction, negatively charged particles in a clockwise direction, and neutral particles not at all. Likewise, stationary particles just remain at rest.

the culmination of decades of work learning how to build bigger and better accelerators.

$E = mc^2$

When the Bevatron created antiprotons, it wasn't because there were antiprotons hidden in the protons and atomic nuclei they were working with. Rather, the collisions brought new particles into existence. In the language of quantum field theory, the waves representing the original particles set up new vibrations in the antiproton field, which we detect as particles.

In order for that to happen, the crucial ingredient is that we have enough energy. The insight that makes particle physics possible is Einstein's famous equation, $E = mc^2$, which tells us that mass is actually a form of energy. In particular, the mass of an object is the minimum energy that object can have; when something is just sitting perfectly still, minding its own business, the amount of energy it possesses is equal to its mass times the speed of light squared. The speed of light is a big number, 186,000 miles per second, but its role here is just to convert units of measurement from mass to energy. Particle physicists like to use units where speed is measured in light-years per year; in that case c is equal to one, and mass and energy become truly interchangeable, $E = m$.

What about when an object is moving? Sometimes discussions of relativity like to talk as if the mass increases when a particle approaches the speed of light, but that's a little misleading. It's better to think of the mass of an object as fixed once and for all, while the energy increases as it goes faster and faster. The mass is the energy that the thing would have if it was not moving, which by definition doesn't change even if it happens to be moving. Indeed, energy grows without limit as you get closer and closer to the speed of light. That's one way of understanding why the speed of light is an absolute limit to how fast things can go—it would take an infinite amount of energy for a massive body to move that fast. (Massless particles, in contrast, always move at exactly the speed of light.) When a particle accelerator pushes protons to higher and higher energies, they are coming closer and closer to the speed of light, never quite getting there.

Through the magic of this simple equation, particle physicists can make heavy particles out of lighter ones. In a collision, the total energy is conserved but not the total mass. Mass is just one form of energy, and energy can be converted from one form to another, as long as its total amount remains constant. When two protons come together at a large velocity, they can convert into heavier particles if their total energy is large enough. We can even collide perfectly massless particles to create massive ones; two photons can smack together to make an electron-positron pair, or two massless gluons can come together to make a Higgs boson, if their combined energy is larger than the Higgs mass. The Higgs boson is more than a hundred times heavier than a proton, which is one of the reasons it's so hard to create.

Particle physicists enjoy using units of measurement that make no sense to the outside world, as it lends an aura of exclusivity to the endeavor. Also, it would be a pain to use one set of units for mass and another for other forms of energy, since they are constantly being converted back and forth to one or the other. Instead, whenever we're faced with an amount of mass, we simply multiply it by the speed of light squared to instantly convert it into an energy. That way we can measure everything in terms of energy, which is much more convenient.

The energy unit favored by particle physicists is the electron volt, or "eV" for short. One eV is the amount of energy it would take to move an electron across one volt of electrical potential. In other words, it takes nine electron volts' worth of energy to move an electron from the

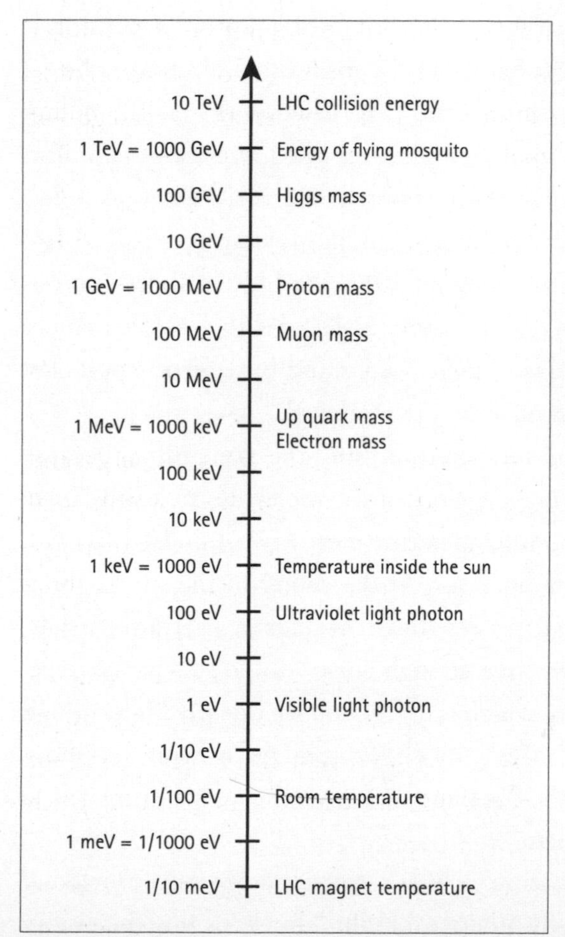

Scale of energies. Particle physicists measure temperature, mass, and energy on the same scale, using electron volts as a basic unit. Common expressions include milli-eV (1/1000 eV), keV (1000 eV), MeV (1 million eV), GeV (1 billion eV), and TeV (1 trillion eV). Some values are approximate.

positive to the negative terminals of a nine-volt battery. It's not that physicists spend a lot of time pushing electrons through batteries, but it's a convenient unit that has become standard in the field.

One electron volt is a tiny bit of energy. The energy of a single photon of visible light is about a couple of electron volts, while the kinetic energy of a flying mosquito is a trillion eV. (It takes many atoms to make a mosquito, so that's very little energy per particle.) The amount of energy you can release by burning a gallon of gasoline is more than 10^{27} eV, while the amount of nutritional energy in a Big Mac (700 calories) is about 10^{25} eV. So a single eV is a small amount of energy indeed.

Since mass is a form of energy, we also measure the masses of elementary particles in electron volts. The mass of a proton or neutron is almost a billion electron volts, while the mass of an electron is half a million eV. The Higgs boson that the LHC discovered is at 125 billion eV. Because one eV is so small, we often use the more convenient unit of GeV, for giga– (1 billion) electron volts. You'll also see keV for kilo– (1,000) electron volts, MeV for mega– (1 million) electron volts, and TeV for tera– (1 trillion) electron volts. In 2012, the LHC collided protons with a total energy of 8 TeV, and the eventual goal is 14 TeV. That's more than enough energy to make Higgs bosons and other exotic particles; the trick is to detect them once they're produced.

We can even measure temperature using the same units, because temperature is just an average energy of the molecules in a substance. From this perspective, room temperature is only two-hundredths of an electron volt, while the temperature at the center of the sun is about 1 keV. When the temperature rises above the mass of a certain particle, that means that collisions have enough energy to create that particle. Even the center of the sun, which is pretty hot, isn't nearly high enough to produce electrons (0.5 MeV), much less protons or neutrons (about 1 GeV each). Back near the Big Bang, however, the temperature was so high that it was no problem.

The easiest way for nature to hide a particle from us is to make it so heavy that we can't easily produce it in the lab. That's why the history of particle accelerators has been one of reaching for higher and higher energies, and why accelerators get names like Bevatron and Tevatron. Reaching unprecedented energies is literally like visiting a place nobody has ever seen.

Energizing Europe

The official name of CERN, the Geneva laboratory where the LHC is located, is the European Organization for Nuclear Research, or in French Organisation Européenne pour la Recherche Nucléaire. You'll notice that the acronym doesn't work in either language. That's because the current "Organization" is a direct descendant of the European Council for Nuclear Research, Conseil Européen pour la Recherche Nucléaire, and everyone agreed that the old abbreviation could stick even after the name was officially changed. Nobody insisted on switching to "OERN."

The council was established in 1954 by a group of twelve countries that sought to reenergize physics in postwar Europe. Since that time, CERN has been at the forefront of research in particle and nuclear physics, and has served as an intellectual center for European science, as well as an important component of Geneva's identity. In the second-largest city in Switzerland, a world center of finance, diplomacy, and watchmaking, one out of sixteen passengers passing through Geneva airport is somehow associated with CERN. When you land there, chances are there's a physicist or two on your airplane.

Like most major particle physics labs, the story of CERN has been one of bigger and better machines reaching ever-higher energies. In 1957, there was the Synchrocyclotron, which accelerated protons to an energy of 0.6 GeV, and in 1959, saw the inauguration of the Proton Synchrotron, which reached energies of 28 GeV. It still operates today, providing beams that are accelerated further by other machines, including the LHC.

A major step forward came in 1971 with the Intersecting Storage Rings (ISR), which attained 62 GeV in total energy. The ISR was a proton *collider* as well as an accelerator. Previous machines had accelerated protons and aimed them at stationary blocks of matter, which are relatively easy targets to hit; the ISR collided beams moving one way with beams moving in the opposite direction. This technique presents a much greater technological challenge but also makes much higher energies accessible; not only does each beam carry energy, but every bit of the energy is now available to make new particles. (In fixed-target experiments, much of the energy goes into providing a push to the target.) Prospects for building a particle collider were studied in the 1950s by Gerard K. O'Neill, an American physicist, who later became more well-known for proposing and advocating human habitats in outer space, and small electron-positron colliders were constructed in Frascati, Italy, in the 1960s by Austrian physicist Bruno Touschek.

The ISR was about one and three-quarters of a mile in circumference. Big, but bigger was yet to come. The Super Proton Synchrotron (SPS), more than four miles in circumference, opened in 1976, and reached energies of 300 GeV. Just a few years later, in a bold move, CERN reengineered the SPS from its original task of accelerating protons to a new configuration in which it collided protons with *antiprotons*. As you might expect, antiprotons are hard to collect and work with. They're not lying around like protons are; you have to make them in lower-energy collisions to start, and then work hard to gather them without their bumping into an ambient proton and annihilating in a flash of light. But if you pull off this trick, you can take advantage of the fact that protons and antiprotons have opposite charges to curve them around in opposite directions but in the same magnetic field. (The LHC collides protons with protons, and therefore has to use two separate beam pipes for the two directions.) Italian physicist Carlo Rubbia used the upgraded SPS in 1983 to discover the W and Z bosons of the weak nuclear force, picking up the Nobel Prize in 1984.

The SPS is still around, and still hard at work. Thanks to upgrades, it now accelerates protons up to 450 GeV. These are handed off to the

LHC, which pushes them to even higher energies. Particle physicists are great believers in recycling.

CERN inaugurated its next great machine in 1989—the Large Electron-Positron collider (LEP). This required yet another new tunnel, this time seventeen miles around and 330 feet underground, stretching across the Swiss-French border. If those numbers sound familiar, they should—the tunnel that was originally built for LEP is the same one in which the LHC now sits. After a successful run, LEP was shut down in 2000, and the machinery was removed to make way for the LHC.

The Large Electron-Positron collider

Protons are hadrons—strongly interacting particles. When you smash two of them together (or a proton and an antiproton), the results are a little unpredictable. What really happens is that one of the quarks or gluons inside the hadron smacks into one of the quarks or gluons in the other, but you don't know the precise energy of either particle to start. A machine that collides electrons and positrons is a very different beast: built for precision, not brute power. When an electron and positron collide, as in LEP, you know exactly what is going on; the results are better suited for delicate measurements of particle properties than for discovering new particles in the first place. If you're playing "Where's Waldo?" particle physics at a hadron collider is like letting your gaze wander over the entire picture looking for a jaunty striped cap; searching at an electron-positron collider is like placing a fine grid over the drawing and painstakingly examining the faces one by one.

LEP was so precise that it was even able to discover the moon. Or, at least, the tides it causes. Each day, the moon's gravitational field tugs at the earth as it rotates underneath. At CERN, this tiny stress caused the total length of the LEP tunnel to stretch and contract by about a millimeter (one-twenty-fifth of an inch) every day. Not such a big deal in a seventeen-mile-long beam pipe, but enough to cause a tiny fluctuation in the energy of the electrons and positrons—one that was easily detectable by the high-precision instruments. After some initial puzzlement at the daily energy variations, the CERN physicists quickly figured out what was going on. (At heart, this way of detecting the moon isn't that different from how astrophysicists detect dark matter in the universe, through its gravitational influence.) LEP was also able to detect the passage of high-speed TGV trains entering Geneva, whose leaking electrical currents were able to measurably disturb the precisely tuned machine.

But the LEP physicists weren't there to detect the moon or trains; they wanted to discover the Higgs boson. And for a while, they thought they had.

After a very successful run making precision measurements of properties of the Standard Model (but not discovering any new particles), LEP was scheduled to be turned off in September 2000 and dismantled to make way for the LHC. Knowing that their machine had only a few months of operation left, the technicians went for broke, using every trick they could think of to boost it to 209 GeV, a higher energy than its design specifications had ever contemplated. If it broke, it was a lame duck accelerator anyway.

As the beams attained these new energies, a team at an experiment named ALEPH led by Sau Lan Wu of the University of Wisconsin–Madison noticed a handful of events that stood out above the rest. Just a few tantalizing hints, but exactly what we might expect if there was a Higgs boson lurking at a mass of 115 GeV, right at the edge of what LEP was capable of seeing. Wu has a number of important results to her name, including sharing the European Physical Society Prize for a 1979 experiment that helped establish the existence of gluons. She was hot on the trail of the Higgs, and wouldn't let this opportunity slip by carelessly.

Ordinarily, a few suggestive events in a particle detector aren't much reason to get excited, even if they look exactly like the Holy Grail that you and your colleagues have been chasing for years. Particle physics is about statistics: For almost anything you can see in a detector, there is more than one way to make it happen, and the whole trick is comparing the rate you should expect against the rate you might get with a new particle. So if a few events are teasing you, just collect more data. The signal will either grow stronger or fade away.

The problem is, you can't collect more data if the lab is going to turn off your accelerator. Wu and other physicists petitioned Luciano Maiani, an Italian physicist who was director general of CERN at the time, to extend the LEP run in order to collect more data. Everyone appreciated the possible significance of the potential discovery, and the enormous regret they would feel if they shut down the machine just before finding the Higgs. You don't often get to see an elementary particle for the first time, especially one this central to our understanding of physics. As physicist Patrick Janot put it at the time, "We are writing a line in the history of mankind." And they also knew they had competition: The Tevatron accelerator at Fermilab, outside Chicago, was also taking aim at the Higgs boson, and might be able to find it at 115 GeV before the LHC could come up to speed. Particle physics relies on international collaboration, but a competitive fire burns inside every scientist.

Maiani, appreciating what was at stake, chose a compromise: LEP would still be shut down, but only after one additional month of running, through October 2000. The Higgs hunters grumbled a little bit but set about collecting more data in search of events that matched what the Higgs should produce. And they found them; just a few, but scattered over the four different experiments running at LEP, not just at the ALEPH detector where Wu's team was working. But they also collected many more "background" events that didn't look like the Higgs at all.

When the run finally came to an end, the total statistical significance of the apparent Higgs events had actually decreased; the signal was being swamped by the background. LEP could have kept running, but that would have meant a serious delay in the schedule for building the LHC, which would have meant both increased costs and more time before the bigger machine would finally come online. As tempting as it was to make one last grab for the brass ring, it was time for LEP to retire and for other accelerators to take up the chase.

SLAC, Brookhaven, Fermilab

While CERN has successfully combined the efforts of many European countries (and more recently, the world) to create a leading physics lab, other facilities have also been responsible for major advances in our understanding of particles and forces. Three labs in the United States, in particular, have helped put together the Standard Model: SLAC at Stanford University in California, Brookhaven on Long Island, and Fermilab outside Chicago.

SLAC originally stood for "Stanford Linear Accelerator Center," but in 2008, the Department of Energy officially changed it to "SLAC Linear Accelerator Center," perhaps because someone in a position of power is fond of infinite recursion. (More plausibly, because Stanford University didn't want the Department of Energy to trademark an acronym containing their name.) Founded in 1962, SLAC holds a unique place in particle physics by hosting a high-energy *linear* accelerator—a straight line rather than a circular ring. The building containing the accelerator is two miles long, the longest building in the United States and third-longest in the world. (The top two are the Great Wall of China and the Ranikot Fort, a nineteenth-century military fortification in Pakistan.) Originally, the accelerator used electrons and slammed them into fixed targets. Starting in the 1980s, it was upgraded to collide electrons with positrons, and eventually a ring was added, using the linear accelerator as a first stage.

SLAC played a key role in the discovery of several particles, including the charm quark and the tau lepton, but undoubtedly its greatest contribution was showing that the very idea of quarks was on the right track. In 1990, the Nobel Prize was awarded to Jerome Friedman and Henry Kendall of MIT and Richard Taylor of SLAC, who in the 1970s used SLAC's electron beam to closely examine the inner structure of protons. The SLAC-MIT team showed that low-energy electrons went right through the protons without much deflection, while high-energy electrons (which you might have expected to go through even more easily) were more likely to careen off at odd angles. Particles with high energies correspond to field vibrations with short wavelengths, and are therefore sensitive to resolve what's going on at very short distances. What the physicists were seeing were very small particles living inside the protons—what we now know as quarks.

Brookhaven National Laboratory was founded in 1947, and has contributed to seven different Nobel Prizes: five in physics and two in chemistry. The muon neutrino, for which Lederman, Schwartz, and Steinberger shared the Nobel, was discovered at Brookhaven. Currently, its main contribution to particle physics comes from the Relativistic Heavy Ion Collider (RHIC), a 2.4-mile-long ring that smashes heavy nuclei together to create the kind of quark-gluon plasma that existed shortly after the Big Bang. Officials from the Guinness Book of World Records have certified that RHIC is responsible for the highest artificially produced temperature of all time: more than 7 million degrees Fahrenheit, or 250,000 times the temperature at the center of the sun. The goal of physics at RHIC is not so much to search for new particles, as to figure out how quarks and gluons behave in these extreme circumstances.

The other major high-energy physics complex is the Fermi National Accelerator Laboratory, or Fermilab for short. Specializing in giant rings that accelerated protons and antiprotons to high energies, Fermilab has been a direct competitor to CERN for much of its existence. It was founded in 1967 under the guidance of Robert Wilson, a polymath scientist and innovative administrator who was renowned among physicists for his creativity and apparent ability to achieve the impossible. Not only did he bring in the new laboratory ahead of schedule and under budget, he designed the main building and personally created many of the sculptures that bring the site to life. When Wilson, who had briefly studied sculpture at the Accademia Belle Arti in Rome, proposed building a thirty-two-foot-tall metal obelisk for the lab, he was told that union regulations required that all welding be done by union members. His response was natural (for him): He joined the welders' union, apprenticed himself to master welder James Forester of the Fermilab Machine Shop, and dutifully followed the appointed course of instruction. The obelisk, constructed by Wilson over lunchtimes and weekends, was installed in 1978 in a reflecting pool outside the main hall.

The pride of Fermilab was the Tevatron, a massive machine that collided protons and antiprotons together at energies of 2,000 GeV. (Remember that "TeV" stands for one "Tera Electron Volt," which is 1 trillion electron volts, or 1,000 GeV.) Completed in 1983, the Tevatron was the highest-energy accelerator in the world until the LHC took the crown in 2009. Its crowning achievement was the discovery of the unusually massive top quark, finally pinning it down in 1995. Gordon Watts of the University of Washington, who was a graduate student working at Fermilab at the time, remembers the moment when the signal climbed above the important "three sigma" threshold (explained in Chapter Nine) for claiming evidence for a new particle:

> We were in one of the big top meetings reviewing all the analyses that were about to go out for one of the conferences. Every analysis was seeing a small excess, but it was so small that it wasn't really meaningful. In fact, they had been doing this for quite some time and we were all used to it—so we basically ignored it. It was the end of one of the normal marathon meetings, the room was packed, I was sitting on the floor in the back, in fact. It was hot, and the room air was . . . umm . . . stuffy (to put it nicely). I think we were about to hear the last talk when one of the people that had gotten there early enough to snag a chair raised his hand . . . "Uh . . . hold it a moment . . . if I do the simplest thing here and add up all the backgrounds and the signals I get over three sigma." There was a silence in the room while everyone went scrambling back through the talks to figure

> out if that was actually correct. Either the spokesperson or the top convener spoke next . . . it was a four-letter word. I think everyone felt the chill go down their spine.

The long sought-after Higgs remained beyond the Tevatron's grasp. With lower energy and luminosity than the LHC, the American machine was always a long shot to win that race. But after LEP turned off, and before the LHC came to life, Fermilab had a window where they could possibly have claimed the first solid evidence of the mysterious particle. In the end, physicists at the Tevatron were able to exclude certain mass ranges for the Higgs, but they couldn't claim any hard indication for its existence.

Facing significant pressure from a gloomy budget situation, as well as the much higher energies of the newly operational LHC, the Tevatron was shut off for good on September 30, 2011, ending the career of the last major high-energy particle collider on U.S. soil. (The Relativistic Heavy Ion Collider at Brookhaven does important work in nuclear physics but doesn't compete in the search for new particles, reaching energies of less than 10 GeV per nucleon.) Whether it will ever have a successor is currently unknown.

The Super Collider

There was supposed to be a successor to the Tevatron, of course: the Superconducting Super Collider, which was endorsed by President Ronald Reagan in 1987 and originally scheduled to begin operation in 1996. The SSC was a grandly ambitious scheme, featuring a brand-new ring fifty-four miles in circumference, colliding protons with 40 TeV of total energy, twenty times higher than the Tevatron. In retrospect, it may have been too ambitious. Support for the project was very high in the early days, when a site for the laboratory had not yet been chosen: Nearly every state's Congressional delegation could imagine they would bring home the massive project for their constituents, and forty-three of the fifty states treated the competition seriously enough to undertake geological and economic surveys. The eventual winner was a site near the sleepy town of Waxahachie, Texas, about thirty miles south of Dallas.

Once the SSC site was selected, enthusiasm for the project immediately dampened among forty-nine of the fifty state delegations in Congress. It was a time of great pressure to bring the federal budget deficit under control, and the SSC cost, high to begin with, had nearly tripled to $12 billion. An additional factor was competition—in the minds of government officials, if not in the minds of scientists—from the International Space Station. The ISS budget was more than $50 billion from NASA, or more than $100 billion if flights of the space shuttle were included in the cost. It did not escape notice that much of the money for this giant project would also end up in Texas, with the Johnson Space Center serving as mission control.

I asked JoAnne Hewett, now a theorist at SLAC, about how she came to accept her current job. She could pinpoint the day precisely: October 21, 1993, the day Congress voted to kill the SSC for good. Hewett had offers from the SSC lab, as well as from SLAC, and was eager to be part of the exciting atmosphere at the new machine under construction. She spent that autumn morning watching Congress on C-SPAN, observing helplessly as the vote went the wrong way. She spent that afternoon in mourning, and then called the director of SLAC to accept their offer. Her career there has been very successful, building new models of particle physics and inventing clever ways to test them against the data—but one can't help but feel wistful about the prospect of actually having such data in hand, earlier and from higher-energy collisions.

I was a newly minted postdoc myself at the time, part of the particle theory group at MIT. I remember a somber meeting we held, inviting the entire Boston-area physics community to get together and talk about what we should do next. Some questions were scientific: Is there any other way to attack the questions the SSC was designed to address? Some were more practical: Should we throw our support behind a serious U.S. investment in the LHC, or should we keep fighting a battle that was already lost? Some were even more practical than that: Is there some way we can offer jobs or temporary positions to the scientists who were thrown out of work by the closing of the SSC lab?

At the time of the SSC's cancellation, $2 billion had already been spent to excavate part of the tunnel and build some of the necessary physical infrastructure. It's hard to pinpoint a single justification for Congress's decision to cancel the project, but a frequent complaint was the reluctance of SSC management to institute proper bureaucratic procedures. A 1994 post-cancellation report from a Congressional staff committee, entitled "Out of Control: Lessons from the Superconducting Super Collider," detailed numerous allegations of mismanagement, including consistently underestimated costs, a failure to carry out mandatory internal reviews, and difficulties in communicating with Congress and the Department of Energy itself. Sometimes the criticisms got silly, as when news stories broke that the laboratory had spent $20,000 on plants (which turned out to include landscaping). The physicists, meanwhile, chafed at what they saw as unnecessary red tape. Roy Schwitters, who was serving as director of the SSC lab, grumbled to a reporter, "Our time and energy are being sapped by bureaucrats and politicians. The SSC is becoming a victim of the revenge of the C students." In retrospect, that might not have been the most politically astute formulation.

Meanwhile, the physicists were fighting among themselves. While particle physics receives a hefty fraction of research dollars and public attention, it is a distinctly minority pursuit within the larger field of physics. Only 7 percent of the membership of the American Physical Society (APS) are members of the Particles and Fields subdivision; others identify as researchers in condensed matter and materials, atomic and molecular physics, optics, astrophysics, plasma physics, fluid dynamics, biophysics, or other specialties. By the late 1980s and early '90s, many in these fields were more than a little irked at all the attention and funding that flowed toward particle physics, and to them the SSC was a symbol of priorities that had gone seriously awry.

Bob Park, executive director of the APS's office of public affairs at the time, said in 1987 that the SSC was "perhaps the most divisive issue ever to confront the physics community." Philip Anderson of Princeton, a respected condensed matter physicist who won the Nobel Prize in 1977, emphasized "the almost complete irrelevance of the results of particle physics not only to real life but to the rest of science," and argued that while the SSC was good science, the money could perhaps be better spent elsewhere. James Krumhansl, a materials scientist from Cornell, who was in line to become president of the APS, believed that the project was siphoning away money from more cost-effective areas of research, and that development of a new particle accelerator should wait until superconducting and magnetic technologies had improved. Particle physicists often hurt their own case among their colleagues by trying to claim advances in other fields, such as magnetic resonance imaging, as spinoffs from accelerator development. As Nicolaas Bloembergen, another Nobel Laureate and APS president, testified in 1991, "As one of the pioneers in the field of magnetic resonance, I can assure you that these are spinoffs of small-scale science."

Somewhat lost in the jostling over bureaucratic control, budget concerns, and disciplinary priority were the larger questions of the meaning of basic research and the importance of discovery for its own sake. In 1993, a new president and many new representatives had been elected, swearing to bring government spending under control. The Berlin Wall had come down and the Soviet Union had collapsed, ending the Cold War and its attendant race for technological superiority. After hitting an apogee with the Manhattan Project during World War II, the influence of high-energy physicists over national policy had been in gradual decline for half a century. Most thoughtful people can agree that the quest to better understand the universe is an important one, but so is finding adequate health care and income security for a nation's citizens. These are difficult priorities to balance against one another in the best of times.

After the SSC was canceled for good, the land and facilities were turned over to the state of Texas, which tried for a long time to sell them to a private owner. It finally succeeded in 2006, when an Arkansas millionaire named Johnnie Bryan Hunt purchased the site for $6.5 million. Hunt's idea was to turn the SSC complex into an unprecedentedly secure data storage facility. The laboratory was equipped with power and telecommunications lines, and the site had been carefully chosen to steer clear of earthquakes and floods. But before the end of the year, the seventy-nine-year-old Hunt had slipped on a patch of ice, damaged his skull, and died. Plans for the data center were scrapped, and the SSC site again lay quiet. As of 2012, the complex has been purchased by a chemical manufacturer that hopes to build a new plant, although the neighbors are strenuously objecting. Whatever the ultimate fate of the SSC laboratory, Waxahachie is not playing a major role in the search for the Higgs boson.

As many predicted, the cancellation of the SSC did not lead to increased funding for other areas of science; indeed, the same enthusiastic Congressional budget-cutters went to work on the rest of the research budget with gusto. There was, however, one acknowledged winner in the unfortunate episode: the Large Hadron Collider. Denied their dream of a flagship machine, U.S. physicists successfully lobbied for an increased role in the LHC. The infusion of money from America helped move the LHC forward in scope, keeping alive the prospect that the Higgs wouldn't stay elusive forever.

FIVE

THE LARGEST MACHINE EVER BUILT

In which we visit the Large Hadron Collider, the triumph of science and technology that has been searching for the Higgs boson.

On September 10, 2008, the Large Hadron Collider came to life. To the cheers of thousands of physicists worldwide, the first protons successfully circulated around the ring. Champagne corks were popped, backs were slapped, speeches were made, and a new era of human discovery had dawned at long last.

Nine days later, it exploded.

Not the entire accelerator, of course. The LHC lives in a circular tunnel 330 feet underground and about seventeen miles in circumference, looping in a circle across the border of Switzerland and France near Geneva. It would take some sort of unimaginable cataclysm for the whole thing to explode. But individual pieces can break.

For the LHC to work, the inside must be kept extremely cold. The machine circulates protons in two different beam pipes: one for moving clockwise, the other counterclockwise, so that the beams can be brought into collision at certain locations where experiments are situated. Both beam pipes travel through superpowerful magnets, whose job it is to curve the protons precisely to stay on their appropriate path.

It's easy to make magnetic fields: Just run electrical current through a loop of wire. To make strong fields, we need a lot of current. But most materials, even high-quality wires, offer some amount of resistance to the flow of current. The problem then is that the wire will start to heat up and ultimately melt. To combat this problem, the wires are cooled to an incredibly low temperature, so that they become superconductors. A superconductor has no resistance at all, so the wires don't rise in temperature when current runs through. The LHC is the largest refrigerator in the world (by a wide margin), and the cooling is achieved via liquid helium, kept at minus 456 degrees Fahrenheit, just 3.4 degrees above absolute zero, the coldest temperature possible.

Here is the worry: If the temperature of the helium rises just a bit, the wires in the magnets cease to be superconductors. When that happens, the huge amount of electrical current running through them meets resistance, and responds by heating up the wires even more. This in turn heats up the helium, and the process runs out of control, with the liquid helium boiling into gas and exploding out of its containers. In operating mode, the LHC magnets are always a hair's breadth away from disaster.

Such a runaway event is known in the trade as a "quench." On September 19, 2008, a seemingly minor electrical problem caused a quench in one magnet, and the troubles quickly spread to other magnets nearby. Lyn Evans, head of the LHC at the time, remembers sitting in the personnel office haggling about some fairly trivial thing, when he got a call on his mobile telling him to come immediately—something looked serious. "When I got over there, even on a computer screen I had never seen such carnage. Red everywhere."

The difficulty was ultimately traced to a faulty connection in a superconducting joint, which caused an electrical arc that pierced a helium vessel. More than fifty magnets had to eventually be replaced out of the 1,232 that work to bend protons around the LHC's ring. The initial reports out of CERN characterized the incident as a "leak," but "explosion" is a more accurate description. More than six tons of liquid helium were released into the tunnel in just a few minutes, and the stresses ripped magnets from where they were bolted to the floor. Safety procedures require that no one is allowed in the LHC tunnel when protons are circulating, but at the time of the incident the beam was actually turned off; fortunately the affected area was empty at the time, and nobody was hurt.

Redoubled efforts

At least nobody was hurt physically. Mentally the damage was severe. Robert Aymar, a French physicist, who was the director general of CERN at the time, put out a press release stating, "Coming immediately after the very successful start of LHC operation on 10 September, this is undoubtedly a psychological blow." After years of hard work, it was deflating to be so close to seeing the LHC running at last, only to be hit by such a jarring setback.

But this is a story with a happy ending. As disappointing as it was, the September 19 explosion galvanized the CERN community around the task of bringing the LHC back to life. Engineers and physicists threw themselves into the task of checking and improving every piece of the machine to make sure it would be able to withstand the unprecedented energies it was expected to tame. This wasn't just a matter of tightening a few screws: Not only did the damaged equipment have to be repaired, but every other piece of the machine had to be brought up to a higher standard of quality. It was slow and demanding work. Not until over a year later did the accelerator seem ready for prime time once again.

Mike Lamont's official title is LHC Machine Coordinator, but a Star Trek fan once described him as the "LHC's Mister Scott." Having spent more than twenty-three years at CERN, it's his responsibility to keep the protons coming in the face of seemingly insurmountable obstacles. Tiny glitches happen all the time, of course, but as the day grew closer when the LHC would finally turn on once again, every bump in the road seemed to be magnified to epic-disaster proportions. During tests on November 3, 2009, temperatures on some of the magnets began to rise due to an electrical malfunction in one of the stations on the surface. Lamont explained to curious reporters that the problem had been traced to a tiny piece of bread on a bus bar. Apparently a passing bird had dropped a bit of baguette from overhead. Lamont and the other engineers quickly patched up the problem, and regular operations were restored—but not before reporters' eyes had grown wide at this bit of news. The *Telegraph* printed a photograph of the CMS detector next to a photograph of a pigeon, with the caption "The Large Hadron Collider (left) and its arch-nemesis (right)."

On November 20, 2009, protons circulated in the LHC for the first time since the accident. Three days later, the beams were brought together to create the first collisions in the machine. A mere seven days after that, the energies had increased to the point where the LHC was the highest-energy accelerator ever built.

Running on an ordinary schedule, the LHC would shut down during the deep of winter in order to save money during those months when electricity in Geneva is most expensive. But in 2009–2010, everyone was impatient, and the crews doubled their efforts to bring the accelerator up to power. The first physics data (as opposed to "commissioning" data used to test the machine) was taken in early 2010. In March 2010, the LHC reached its provisional energy goal (half of the ultimate target), setting a record for high-energy particle collisions in the process. Champagne flowed once again.

In retrospect, the accident in September 2008 helped the physicists and technicians at the LHC understand their machine much better, and as a result, the physics runs beginning in 2010 were stories of essentially uninterrupted progress. Given that operations didn't start in earnest until that year, it came as a surprise to almost everybody that the experiments collected and analyzed enough data to discover the Higgs by July 2012. It's as if you purchased an expensive car that breaks down almost immediately, and you have to spend a while combating some pesky maintenance problems. But once you finally get it on the road and hit the accelerator, the performance takes your breath away.

The Large Hadron Collider is Big Science at its biggest. The number of moving parts—human as well as mechanical—can sometimes be intimidating, or even depressing. In the words of Nobel Laureate Jack Steinberger, "The LHC is a symbol of just how difficult it is these days to make any progress. What a difference when compared to my thesis days, sixty-five years ago, when I, singlehandedly, in half a year, could do an experiment which marked an interesting step forward." The LHC is the largest and most complicated machine ever built by human beings, and sometimes it's a surprise that it works at all.

But it does work—spectacularly well. Over and over again, physicists I talked to while writing this book spoke of the awe-inspiring immensity of the operation, but also about how CERN could serve as a model for large-scale international collaboration. Experimentalist Joe Incandela said, "What's amazing to me is that we have people from seventy countries around the world working—together. Palestinians and Israelis working side by side, Iranians and Iraqi scientists work together—such collaborations in the pursuit of Big Science shouldn't be overlooked." Joe Lykken, an American theoretical physicist at Fermilab, wistfully mused, "If only the United Nations could work like CERN, the world would be a much better place."

If you believe that it's a worthwhile task to pursue particles like the Higgs boson that require a huge amount of energy to create, Big Science is the only way to go. There is a tremendous amount of fantastic research to be done that can be tackled with relatively inexpensive tabletop experiments, but discovering new massive particles isn't in that category. Right now the LHC is the only game in town, and its performance is a testimony to human ingenuity and perseverance.

Years of planning

The LHC is a marvel of planning and design. Physicists at CERN had been thinking about a giant proton collider for a while, but the first "official" discussions about what would eventually become the LHC were held at a workshop in Lausanne, Switzerland, in March 1984. The planners knew that the United States was contemplating what would eventually become the Superconducting Super Collider, so they needed to decide whether a European competitor was a sensible use of scarce resources. (They didn't know, of course, that the SSC would eventually be canceled.) Unlike the SSC, which started from scratch building a new facility, the LHC would be limited in scope by the need to fit inside the already-constructed LEP tunnel. As a result, the target energy was set at 14 TeV, barely more than one-third of the 40 TeV target for the SSC. But the LHC would be able to produce more collisions per second, and was less expensive—and maybe all the good physics would be accessible at 14 TeV, rendering the higher energy of the SSC irrelevant.

Much of the impetus for moving forward with the LHC came from Italian physicist Carlo Rubbia, a brash and influential experimentalist who had collected a Nobel Prize in 1984 for his discovery of the W and Z bosons. Rubbia is a larger-than-life figure, as well-known for his forceful personality as for his accomplishments as a scientist (which are considerable). It was he who cajoled CERN into building the first proton-antiproton collider in 1981, a concept that would later be adopted by Fermilab's Tevatron. (With the LHC we are back to colliding protons on protons, as it is too difficult to make a sufficient number of antiprotons to create the sought-after number of collisions.)

First as the chair of CERN's Long Range Planning Committee, and later as director general of the lab from 1989 to 1993, Rubbia pushed strongly for the LHC at a time when LEP wasn't yet finished and the United States was thought to be moving forward with the SSC. Europe was facing its own budgetary woes, especially in Germany, where the costs of reunification were running high. Rubbia was eventually able to convince the European governments that a hadron collider was the logical next step for the lab, regardless of what other countries might be doing. It wasn't until 1991 that the CERN council adopted a resolution to officially study the LHC proposal, and not until December 1994 (after the SSC was canceled) that the project was finally approved. Lyn Evans was appointed director of the LHC, and the massive task of moving from idea to reality began in earnest.

The architect

In a project stretching over so many years, involving so many people and countries, and with such an intimidating number of significant subprojects, it would be unfair to give too much credit to any single person, downplaying the role of so many others. Nevertheless, if any individual is to be mentioned as having built the LHC, it would be Lyn Evans.

Evans comes across as an unassuming man, gray-haired and distinguished-looking but informal. Born to a mining family in Wales, his first love was chemistry; he took special joy making explosives, perhaps a fitting start for someone who would one day engineer the highest-energy particle collisions humans have ever achieved. In university he switched to physics because "physics was more interesting, and easier." When the LHC project was approved, CERN needed someone with enough experience to manage the job, but young and energetic enough to see it through to completion. Evans was handed the daunting task of squeezing the highest possible physics return out of a machine with a fixed size, a limited budget, and an array of technological challenges that were unique in the history of experimental science. It was Evans who figured out how to take the original schematic plans for the LHC and modify them into a design that was compatible with financial realities.

During the progress of an engineering project of this magnitude, unanticipated roadblocks are going to pop up. While the LHC already had a waiting tunnel courtesy of LEP, new caverns had to be excavated for the four large experiments that would be used to measure the outcomes of the collisions. The CMS experiment sits on the far side of the ring from the main CERN site, near the town of Cessy, on the French side of the border. When workers set about digging a hole for the new experiment, they made an unanticipated discovery: the ruins of a fourth-century Roman villa. Jewelry and coins from what are today England, France, and Italy were found at the site. Fascinating for archaeologists, but a critical delay for the physicists; construction stopped for six months while the ruins were examined.

That was far from the end of it. The location of the CMS cavern turns out to sit beneath an underground river. The flowing water isn't enough to disturb the experiment itself, but it posed problems for the excavation process. The construction team came up with a very physics-like solution: They sank pipes into the ground and filled them with liquid nitrogen, freezing the water into ice and giving the diggers solid ground to work with. "That was quite exciting," Evans observed.

Evans, and the many other physicists and CERN staff working on the LHC, persevered. Apart from technical problems, skittish governments were constantly threatening to cut their contribution to CERN. At the highest levels, particle physics requires as much diplomacy and political savvy as scientific and technical know-how. A major step forward was achieved in 1997, when the United States agreed to contribute $2 billion to the project. All of the official member states of CERN are European: Austria, Belgium, Bulgaria, the Czech Republic, Denmark, Finland, France, Germany, Greece, Hungary, Italy, the Netherlands, Norway, Poland, Portugal, the Slovak Republic, Spain, Sweden, Switzerland, and the United Kingdom. The United States (along with India, Japan, Russia, and Turkey) is an "observer" state, allowed to participate in physics operations and attend meetings of the CERN council, but not to officially contribute to setting policy. Many other countries have agreements allowing their scientists to work at CERN. But the United

States is the gorilla in the room, and securing a major commitment to the success of the LHC played a significant role in its development, as did earlier commitments from Japan and Russia. More than a thousand American physicists were soon working on the LHC.

Evans has a naturally easygoing style, and is more comfortable getting his hands dirty with a piece of equipment than demanding that underlings keep careful records of ongoing progress. While construction on the LHC proceeded according to plan, tiny cost overruns gradually accumulated. Matters came to a head in 2001, when it was realized that the accelerator was approximately 20 percent over budget. Against Evans's judgment, Director General Luciano Maiani revealed the overrun in an open CERN council meeting, directly requesting that the member states pony up to cover the extra cost.

They were not happy. Robert Aymar, who would follow Maiani as director general in 2004, was instructed by the CERN council to undertake a close look at the management of their flagship machine. Some questioned whether Evans was the right man for the bureaucratic task, and whether a sterner hand wasn't required. But Aymar understood that Evans's unique understanding of the LHC was far more valuable than any looseness of style, and he was kept on as director of the project. Evans would later characterize this time as a low point in his work on the LHC. "I really got a grilling," he said. "That was the worst year of all."

On the September 19 incident after the machine had started up, Evans reflected, "This was the last circuit on the last sector, so it was a bitch. Fortunately, I've had some hard problems in the past."

Accelerating particles

In a game of tetherball, one end of a rope is attached to a volleyball and the other to the top of a pole. Two combatants stand on opposite sides of the pole, whacking at the ball in an attempt to wind the rope around the pole. Now imagine there is just a single player, and that the rope can revolve freely around the top of the pole rather than get twisted up. On each revolution, the player pushes the ball in the same direction, nudging it toward ever-faster speeds.

In a nutshell, that's the basic idea behind a particle accelerator such as the LHC. The role of the volleyball is played by a bunch of protons. The role of the rope that keeps the ball moving in a circle is played by strong magnetic fields that curve the protons around the ring. And the role of the player hitting the ball is played by an electric field that pushes the protons to increase their speed on each revolution.

Protons are extremely small by everyday standards, about one ten-trillionth of an inch across. You can't just pick one up and throw it or whack at it with your hand as it passes by. To accelerate the protons in the LHC, a voltage generator creates a rapidly varying electric field that switches its direction as the protons pass, about 400 million times a second. The switching is timed very precisely, so that any given proton always sees an electric field pointing in the same direction as it traverses through the cavity, swiftly imparting greater velocity. This boost happens only at one point along the ring; most of the effort over the twenty-seven-mile course is simply spent keeping the protons turning in the appropriate direction, not making them go faster.

When the LHC is going full steam, there are a total of about 500 trillion protons circulating in two beams, one moving clockwise and the other counterclockwise around the ring. (Numbers are approximate because the machine's performance gradually improves over time.) That's a lot of protons, but it's still a tiny number compared with any everyday object. All of the protons in the LHC come from a single unassuming canister of hydrogen, which looks for all the world like a fire extinguisher. A molecule of hydrogen has two atoms, each with just one proton and one electron. A bit of molecular hydrogen is extracted from the canister, then zapped with electricity to strip off the electrons, and the protons are sent on their way. Lyn Evans, who had been trained in fusion science rather than particle physics, got his start at CERN working on just such a process. There are about 10^{27} hydrogen atoms in the canister, which is enough to keep the LHC running for about a billion years. Protons are not a rare resource.

Protons aren't continuously injected into the LHC; they come in the form of a "fill," which is added all at once, and maintained for about ten hours (or until the beam degrades for some reason). The protons are moved with utmost care through a series of preliminary accelerators before they finally enter the main ring. There is no room for sloppiness. The protons in the two circulating beams aren't spread uniformly—they are grouped into thousands of "bunches" per beam, with more than 100 billion protons per bunch. The bunches are about an inch long, twenty-three feet apart, and focused into a very thin needle. The beam is about one twenty-fifth of an inch across while traveling around the ring—about the width of the lead in a pencil—and gets further concentrated down to one one-thousandth of an inch as the bunches enter a detector in order to collide. Protons all have an equal positive electric charge, so their natural tendency is to push apart from one another, and keeping the beam under control is a major task.

Besides the energy of the colliding particles, the other important quantity in an accelerator is the luminosity, which is a way of measuring how many particles are involved. You might think we could just count the number of particles zooming around, but what really matters is the number of collisions, and a lot of particles only lead to a lot of collisions if the beam is focused very tightly. During 2010, the priority was on shaking down the machine and checking that everything was in working order, so the luminosity wasn't very high. By 2011, the kinks were largely worked out, and they collected about one hundred times as many collisions as in the previous year. In 2012, the success continued, and during the first half of the year they had more collisions than in all of 2011. That blaze of data is what enabled the sooner-than-anticipated discovery of the Higgs.

Speed and energy

The LHC's protons have a lot of energy because they are moving fast—very close to the speed of light. Every massive object, whether a person or a car or a proton, has some amount of energy when it's sitting still, from Einstein's formula $E = mc^2$, and an additional "kinectic" energy that depends on how fast it's moving. In the everyday world, the energy of motion is much, much less than the energy an object has even at rest, just because everyday velocities are much, much less than the speed of light. The fastest airplane in the world is a NASA experimental craft called the X-43, which reaches speeds of up to seven thousand miles per hour; at that velocity, the plane's energy of motion adds only one ten-billionth of its energy at rest.

Protons in the LHC move quite a bit faster than the X-43. During its first 2009–2011 run, they were traveling at 99.999996 percent of the speed of light, or 670,616,603 miles per hour. At those velocities, the energy of motion is much greater than the energy at rest. The rest energy of a proton is just a shade under one GeV. The first run of the LHC featured protons with 3,500 GeV of energy each, or 3.5 TeV for short, so that when two of them collided there was a total of 7 TeV of energy to go around. The 2012 run collided protons with a total of 8 TeV of energy, and the eventual goal is to reach 14 TeV. Fermilab's Tevatron, by contrast, maxed out at about 2 TeV of total energy.

At velocities this close to the speed of light, the theory of relativity becomes crucially important. Relativity teaches us that space and time change at high velocities: Time slows down compared to clocks at rest, and lengths get contracted along the direction of motion. As a consequence, the seventeen-mile trip around the ring would appear like a much shorter journey to one of the high-energy protons, if protons noticed such things. At 4 TeV, a proton would perceive one trip around the ring to extend only twenty-one feet. Once they get up to 7 TeV per proton, it will be only twelve feet.

How much energy is a TeV? Not that much—about equal to the energy of motion of a mosquito in flight, not something you would notice if it bumped into you. The amazing thing is not that 4 TeV (or whatever) is so much energy, it's that all that energy is packed into a single proton. And remember that there are 500 trillion protons zooming around inside the LHC. If we take the beam as a whole, now we're talking serious energy—about the same energy of motion as an onrushing locomotive engine. You wouldn't want to get in the way.

Or would you? While the protons in the LHC pack a considerable punch, they are collimated into a very fine beam. Maybe most of them would pass right through you?

Yes and no. Nobody has ever stuck any body parts into the LHC beam, nor could they possibly; it's tightly sealed in a vacuum tube, inaccessible to meddling humans. But in 1978, an unfortunate Soviet scientist named Anatoli Bugorski did manage to take a high-energy particle beam right in the face. (Safety standards at the U-70 Synchrotron in Protvino, Russia, were a bit more lax than they are at CERN.) The beam that hit Bugorski consisted of 76 GeV protons—much less than the LHC but still considerable. He was not instantly killed—indeed, he's still alive today. Bugorski later testified that he saw a flash of light, "brighter than a thousand suns," but he reportedly didn't feel pain. He received significant radiation scarring, lost hearing in his left ear, and became paralyzed on the left side of his face; he still suffers from occasional seizures. But he survived without noticeable mental impairment, went on to finish his PhD, and continued to work at the accelerator complex for years afterward. Still, experts recommend avoiding beams of high-energy protons.

The reason why Bugorski's head was not blown to smithereens is that many of the protons did indeed simply pass through him. But at the LHC, it is often necessary to "dump" a fill, which means putting the entire energy of the beam somewhere. (If you could just slow the protons down they would harmlessly dissipate, but that's not practical.) Another way of thinking of that total energy is that it adds up to about

175 pounds of TNT. And it all has to go somewhere, every ten hours or so at the end of a fill.

Experiments have demonstrated that the full brunt of the LHC beam would be sufficient to melt a ton of copper. You certainly don't want it careening randomly into your finely tuned experimental apparatus. Instead, a dumped beam is deflected and diffused away from the normal beam line by special magnets, after which it travels half a mile before landing in a special graphite "dump block." The graphite material is especially good at spreading the energy and not melting in the process, despite reaching temperatures of 1,400 degrees Fahrenheit. There are about ten tons of graphite in total, all of which are encased in one thousand tons of steel and concrete shielding. Give it a few hours to cool down, and you're ready for the next beam dump.

Mighty magnets

We think of the LHC as a giant circular ring seventeen miles around, but it's actually more like a curvy octagon, with the ring divided into octants. There are eight arcs, each over a mile and a half long, and the arcs are connected by straight sections about a third of a mile long. If you were to visit one of the arcs in the LHC tunnel, you'd see a series of big blue tubes stretching in either direction—the "dipole magnets" that guide the protons as they pass down the beam pipe. There are 154 of these tubes along each arc, each of them fifty feet long and weighing over thirty tons. The inside of each tube is mostly taken up by an ultracold superconducting magnet, and in the very center are two narrow beam pipes through which the protons move—one with particles moving clockwise, the other counterclockwise.

If a charged particle like a proton sits stationary in a magnetic field, it doesn't feel any force at all; it can happily stay there at rest. But when a moving charged particle passes through a magnetic field, it gets deflected from a straight line. (Neutral particles would pass right through, unaffected.) Remember that the LHC beam has the energy of a moving train; we need such incredibly powerful magnets simply because it's not easy to bend the protons in a tight curve.

The LHC magnets are as strong as they can be, to allow for the highest possible proton energies in a tunnel of fixed size. The earth has a magnetic field, which helps your compass tell the difference between north and south; the field inside one of the LHC dipoles is about 100,000 times stronger than the earth's. So strong, in fact, that ordinary materials aren't up to the job, and superconductors are required. The magnets contain almost five thousand miles of wound cable, made from a superconducting compound of niobium and titanium, cooled to ultralow temperatures by 120 tons of liquid helium. The inside of the LHC is actually colder than outer space: the magnet temperatures are lower than that of the cosmic background radiation left over from the Big Bang.

Temperature isn't the only criterion by which the LHC compares favorably with outer space. The interior of the beam pipes, the tubes through which the protons actually travel, must be kept as empty as absolutely possible; if they were filled with air, the protons would constantly be running into the air molecules, destroying the beam. So the beam pipes are kept in a very strict vacuum, so much so that the pressure inside the pipe is about the same as the atmospheric pressure on the moon.

Before the machine was started for the first time, the LHC team worried about whether they had made the beam pipe as empty as required. When the Tevatron started up at Fermilab in 1983, the first attempts to circulate protons quickly fizzled out; the culprit was ultimately discovered to be a tiny piece of tissue clogging the pipe. But how do you easily check seventeen miles of accelerator? The beam pipes themselves are only about an inch across, which led to an ingenious idea: Technicians made a kind of "Ping-Pong ball" from impact-resistant polycarbonate, stuck a radio transmitter inside, and sent it rolling down the pipe. If the ball got stuck, technicians could track the transmissions and figure out where it had stopped. It was a neat idea, and someone was probably disappointed when the balls rolled through unscathed, giving the beam pipes a clean bill of health.

The LHC magnets are the biggest and bulkiest parts of the machine, and represent an extraordinary triumph of technological innovation as well as international collaboration. That level of precision doesn't come cheaply. It's hard to put an exact cost on the LHC, because many expenses go into the upkeep of the lab in general, but a figure around $9 billion gives a good feeling for the total budget. In the words of physicist Gian Giudice, "When expressed in euros per kilogram, the price of the LHC dipoles—the most expensive part of the accelerator—is the same as Swiss chocolate. Were the LHC built of chocolate, it would cost about the same."

Chocolate might not sound very expensive; after all, we eat it. But usually not seventeen miles' worth of the very best. It all adds up.

Passing the torch

Lyn Evans officially retired from CERN in 2010, after the machine was successfully up and running. He had first joined the lab in 1969, giving him more than four decades of experience, serving through ten different directors general. Back in 1981, he, Carlo Rubbia, and Sergio Cittolin, an Italian physicist with a penchant for decorating lab notebooks with Leonardo da Vinci–style sketches, were the only three people in the control room at 4:15 a.m., when they turned on the upgraded Super Proton Synchrotron and witnessed the first proton-antiproton collisions inside a particle accelerator.

Quite a difference from September 10, 2008, when the inauguration of the LHC was an international event witnessed by hundreds of people live and thousands more watching by Internet feeds around the world. On that day, Evans served as master of ceremonies in an LHC control room packed with news media, famous scientists, and visiting dignitaries. Drawing out the suspense, they didn't simply push protons all the way around the ring, but opened up the eight sectors one by one. After the first seven sectors had been successfully navigated, Evans counted down as they prepared the protons to make a full circle of the ring. At the appointed moment, two dots flashed on a gray computer screen, indicating that the beam had both successfully left and arrived back at the same point. The room broke into applause, and a new era in particle physics had begun.

Physicists rarely retire in the conventional sense, and for Evans the new phase of his life will involve joining the CMS experiment at the LHC and helping to plan the next generation of accelerators. After the seminars announcing the discovery of the Higgs, he took a moment to muse on what it had felt like. "I went to the CMS summer party the other day, and there were about five hundred people there. When I see all these young people, I suddenly realize what a weight has been on my shoulders. I mean, how many people are relying on this machine to perform?"

Now that the machine is zooming along, CERN hopes that it will continue on for decades to come. It took more than a year to recover from the September 2008 setback, but since coming back to life the machine has performed splendidly. Running at 7 TeV of total energy through 2010 and 2011, then at 8 TeV in 2012, enabled the discovery of the Higgs boson or something very much like it. Still, the ultimate goal is to hit 14 TeV, and to achieve that will require shutting down for two years while equipment is tested and improved. The shutdown was originally planned to begin in late 2012, but after the discovery the CERN council decided to keep it running at 8 TeV for another few months. It's a natural reaction; whenever you get a new toy, you want to play with it right now.

SIX

WISDOM THROUGH SMASHING

In which we learn how to discover new particles by colliding other particles at enormous speeds, and watching what happens.

As a child, I was fascinated by all kinds of science, but only two subjects really captured my attention: theoretical physics and dinosaurs. (When I was twelve, I didn't know the word "paleontology.") I flirted with other sciences, but the relationships never went very far. My junior chemistry set was fun, mostly because I could set things on fire, but I was never entranced by the thrill of creating new compounds in carefully controlled conditions.

But dinosaurs! There was true romance. My grandfather would take my brother and me to the New Jersey State Museum in Trenton, where we would skip right past the boring artifacts and history exhibitions to gawk at the ominously looming skeletons. I never seriously considered paleontology as a career, but every scientist I know secretly agrees that dinosaurs are the epitome of cool.

Which is why I was thrilled when, as a grown-up faculty member at the University of Chicago, I got the chance to go on a dinosaur expedition. Most paleontological outings do just fine without bringing physicists along, but this expedition was organized by Project Exploration, a nonprofit outfit devoted to bringing science to children and underrepresented minorities. It was a special event for friends of the organization, and I was brought along to provide a different kind of science outreach. Didn't really matter to me—they could have said I was brought along to wash dishes, all I cared was that I was going to dig up dinosaur bones.

And dig we did, in a region of the Morrison geological formation near Shell, Wyoming (population approx. 50). The Morrison is chock-full of fossils from the Jurassic, and we whiled away the daylight heat cheerfully digging up specimens of *Camarasaurus*, *Triceratops*, and *Stegosaurus*. "Digging up" might give an exaggerated sense of the accomplishments of our largely amateur crew; mostly we made some progress on sites that would eventually be covered up and left for another trip to finish.

This experience taught me a great deal—primarily that theoretical physics is a cushier job than paleontology. However, it also answered a question that had been bugging me for years: How do you tell the difference between a piece of fossilized bone and the rock matrix that surrounds it? Over the course of millions of years, the original skeleton absorbs mineral from the rock nearby, until eventually it is more rock than bone. How do you distinguish one from the other?

The answer: very carefully. There are tricks, of course, honed over the course of an expert paleontologist's career; subtle gradations of color and texture that elude the notice of the uninitiated. Bring a group of amateurs to a dinosaur fossil site, and far and away the most common question you will hear is "Is this a bone?" But there is a right answer, and the experts can (almost) always provide it.

While the experience of digging up dinosaurs is worlds away from the everyday life of a theoretical physicist, the similarities with *experimental* particle physics are evident. We speak informally of "seeing a Higgs boson" at the Large Hadron Collider, but the reality isn't that simple. We never see Higgs bosons, nor do we ever expect to, any more than we expect to see dinosaurs walking down the street. The Higgs is very short-lived—one will survive for one ten-billionth of a trillionth of a second, far too short to be captured directly, even by the technological marvels that are the LHC experiments. (A bottom quark, with a lifetime of one-trillionth of a second, is just barely at the verge of being discernible; the Higgs lifetime is one ten-billionth of that.)

What we expect to find is *evidence* for the Higgs boson, in the form of other particles that are created when it decays. Fossils, if you will.

The last chapter talked about the LHC accelerator itself, which zips hundreds of billions of protons on circular paths around a tunnel underneath the suburbs of Geneva. In this chapter we address the experiments—the massive detectors located at particular installations around the ring, where protons are brought into collision in a rapid-fire series of interactions. In the data from some individual event, we might possibly find two sprays of strongly interacting particles, as well as a high-energy muon-antimuon pair. So did all that come from the decay of a Higgs, or from something else? The task of identifying these fossils correctly is a combination of science, technology, and black magic that lies at the heart of the hunt for the Higgs.

Identifying particles

Particle physics is a detective story. Arriving at the scene of a crime, most detectives aren't lucky enough to be greeted by clear videotape footage of the perpetrator committing the act or unimpeachable eyewitness testimony or a signed confession. More likely, there are a few haphazard clues—partial fingerprint here, tiny DNA sample there. The tricky part of the job is piecing together those clues to put together a unique story of the crime.

Likewise, when experimental particle physicists are analyzing the results from a collider, they don't expect to see a little sign attached to a particle saying, "I'm the Higgs boson!" The Higgs will decay quickly into other particles, so we have to have a good idea of what we expect those particles to be—that's a job for the theorists. Then we collide protons together and watch what comes out. Most of the volume inside a particle detector is filled with material in which particles leave telltale tracks as they pass through, the particle-physics analogue of someone's muddy footprints at the crime scene. Of course, not all footprints are muddy: Particles like neutrinos, which don't interact through either electromagnetism or the strong force, don't leave much of a trail at all, and we have to be more clever.

Sadly, the tracks we do see don't come with labels reading, "I'm a muon, and I'm moving at 0.958 the speed of light!" either. We have to deduce what particles emerged from the collision, and what that means for the processes that made it happen. We need to know whether this muon was produced by the decay of the Higgs, the decay of a Z boson, or a number of other suspects. And the particles themselves aren't going to confess.

The good news is that the total number of particles in the Standard Model is relatively manageable, so we don't have too many suspects to consider. We're more like the sheriff of Mayberry than a detective in Manhattan. We have six quarks, six leptons, and a handful of bosons: photons, gluons, Ws, Zs, and the Higgs itself. (Gravitons are essentially never produced, just because gravity is so weak.) If we're able to determine the mass, charge, and whether it feels the strong interaction, we can basically identify it uniquely. So that's the task of the experimentalist: Keep track as precisely as possible of the particles emerging from a collision, and determine their masses, charges, and interactions. That lets us reproduce the underlying process that caused all the excitement.

It's pretty easy to judge whether a particle feels the strong interactions, for the happy reason that those interactions are really strong. Quarks and gluons leave completely different signatures in a detector from the ones leptons and photons do. They are quickly confined into different kinds of hadrons—either combinations of three quarks, known as "baryons," or pairs of one quark and one antiquark, known as "mesons." These hadrons readily bump into atomic nuclei, making them easy to pick out. In fact, when you produce a single high-energy quark or gluon, the strong interactions usually cause it to fragment into a whole spray of hadrons, known as a "jet." That makes it very easy to see that you've made a quark or gluon, but a little trickier to measure its precise properties.

Likewise, it's pretty easy to figure out the electric charge on a particle, thanks to the magic of magnetic fields. Just as the LHC tunnel is filled with powerful magnets that nudge protons around the circular beam pipe, the LHC detectors are suffused with magnetic fields that push different particles in different directions, helping us to identify what they are. If a moving particle is deflected in one direction, it has a positive charge; if it's deflected the other way, it has a negative charge. Moving in a straight line means it's neutral.

Experiments around the ring

When Carl Anderson discovered the positron back in the 1930s, his cloud chamber was about five feet across and weighed two tons. The experiments at the LHC are a bit larger. The two biggest experiments, the general-purpose behemoths that will be looking for the Higgs, are called ATLAS (A Toroidal LHC ApparatuS) and CMS (Compact Muon Solenoid). They are located on opposite sides of the ring, with ATLAS sitting near the main CERN site and CMS over the border in France. The word "compact" is relative, of course; CMS is nearly 70 feet long, and weighs about 13,800 tons. ATLAS is larger in size but also more lightweight, coming in at 140 feet long and 7,700 tons. That's the kind of scale you need to dig down to where we hope the Higgs is lurking.

The LHC also features five other experiments: two medium-size ones, ALICE and LHCb, and three small ones, TOTEM, LHCf, and MoEDAL. LHCb specializes in studying the decays of bottom quarks, which are useful for doing precision measurements. ALICE (A Large Ion Collider Experiment) is constructed to study the collisions of heavy nuclei rather than protons, re-creating the quark-gluon plasma that filled the universe shortly after the Big Bang. That's why it's the Large

"Hadron" Collider, rather than the Large "Proton" Collider—one month a year, the LHC accelerates and collides lead ions instead of protons. TOTEM (TOTal Elastic and diffractive cross-section Measurement), located near CMS, studies the inner structure of protons and will perform precise measurements of the probability they will interact with one another. LHCf ("f" for "forward") uses splashes from collisions to study the conditions under which cosmic rays propagate through the atmosphere. It's located near ATLAS, and is much smaller than the other experiments: two detectors, each less than three feet across. MoEDAL (Monopole and Exotics Detector At the LHC) carries out specialized searches for very unusual particles.

It's the two big experiments, ATLAS and CMS, that have been leading the hunt for the Higgs boson. Unlike the smaller experiments, which are designed with quite specific purposes in mind, these two detectors are made simply to watch protons smash together and do the best possible job at determining what comes out of the collisions. They approached the design challenges somewhat differently, but their capabilities end up being comparable. Needless to say, having two experiments is infinitely preferable to having only one—any dramatic and surprising discovery made by one of the detectors won't be taken seriously until the other one verifies the finding.

It's hard to get a feeling for the immensity of these machines without visiting them in person, which I was able to do while they were still under construction. A person is so small compared with CMS or ATLAS that you usually don't notice them in photographs until someone points them out. Standing next to either detector, you are struck not only by their size but by their complexity. Every piece counts—and, given the international nature of the collaborations, it's quite likely that two neighboring pieces were constructed in laboratories on opposite ends of the globe.

While CMS might not be "compact" in the sense of "small," it is certainly compact in the sense of tightly packed together. It was stuck with the less-desirable location, a good car drive from the CERN buildings, because geological analysis revealed that the nearer location was the only one that could handle ATLAS's greater size. CMS is an extremely dense collection of metal, crystal, and wire. The main magnets, the most powerful of their kind ever built, were constrained to be no more than twenty-three feet across, for very prosaic reasons: Anything larger would have been unable to fit on a truck that could make the trip through the streets of Cessy, the tiny French town where the experiment is located. (The *Wikipedia* page for Cessy, clearly written by physicists working at CMS, advises getting lunch at a certain local pizzeria but warns that "the service can be quite leisurely, so don't go if you are in a hurry.") Financial constraints, as well as logistical ones, played a crucial role in design and construction; the brass in the giant cylindrical end caps on either side of the detector was salvaged from Russian artillery shells. A crucial part of the detector is a set of 78,000 lead tungstate crystals, grown in Russia and China over a period of ten years, taking about two days to artificially grow each crystal.

It is ATLAS, however, that is more likely to be depicted in popular photos of the LHC. The reason is simple: It looks like an alien spaceship. The distinguishing features of the detector are the eight giant toroidal magnets that give the experiment its name. You might not recognize an ATLAS magnet as a "torus," which is the shape of a doughnut, whereas the magnets are tubes that are vaguely rectangular with rounded corners. But physicists learn from topologists, mathematicians who care about general features and not specific shapes, so to them a "torus" is any cylinder that loops back on itself. The ATLAS toroids create a gigantic region of high magnetic field, useful for tracking high-energy muons created in the inner regions of the detector. When the magnets are turned on, the total amount of energy stored in them is more than 1 billion joules—the equivalent of about five hundred pounds of TNT. Fortunately, there's no way for that energy to be released in an explosion. (Energy isn't dangerous unless there's a way to release it. The rest energy in an apple is equivalent to about a million tons of TNT, but it's not really dangerous unless you bring it in contact with an anti-apple.)

The tremendous physical size of ATLAS and CMS is matched by the size of the collaborations that built and run them. The two groups of people are roughly similar: more than three thousand scientists each, representing more than 170 institutions from thirty-eight different countries. The whole group never gets together in the same place at the same time, but an endless stream of emails and videoconferences keeps the different subgroups in constant contact.

If there are two big collaborations carrying out very similar experiments to look for essentially the same phenomena, does that mean they are in competition with each other? Do you really need to ask? There are extremely high-stakes and serious—although mostly respectful—competition between the two experiments, as both race to make new discoveries. And with teams that large, there is a great deal of competition *within* each experiment, as different physicists jockey for positions of power, as well as debate the relative merits of different ways of analyzing the data.

But the system works. It might lead to some scientists with frayed nerves and a shortage of sleep, but the friendly rivalry between and within the experimental groups leads to topflight science. Everyone wants to be first, but nobody wants to be wrong, and if you're sloppy, someone else will quickly figure it out. The well-matched capability of the CMS and ATLAS teams is one of the strongest reasons we will have to trust any results they both agree on—including the discovery of the Higgs boson.

Colliding protons

The task of these mammoth experiments is to figure out what happens when two protons collide at enormously high energies. A proton is not an infinitesimally small particle nor an undifferentiated blob of proton-stuff; it's made of many strongly interacting constituents. We often say, "A proton is made of three quarks," but that's a bit sloppy. The two up quarks and one down quark that make a proton a proton are called the "valence quarks." In addition to those valence quarks, quantum mechanics predicts that there are a large number of "virtual particles" constantly popping in and out of existence: gluons as well as quark-antiquark pairs. It's the energy contained in these virtual particles that explains why protons are so much heavier than the valence quarks that give them their identity. It's hard to give a precise count of how many virtual particles there are, as the number depends on how closely we look. (That's quantum mechanics for you.) But the number of valence quarks remains fixed. If you add up the total number of up quarks inside a proton at any one moment, it would always be exactly two more than the number of antiup quarks; likewise, the number of down quarks is always one more than the number of antidowns.

Basically a proton is a floppy bag of quarks, antiquarks, and gluons, moving around the LHC beam pipe near the speed of light. Richard Feynman dubbed all these constituent particles "partons." According to relativity, objects moving near light speed are contracted along their direction of motion. So two protons colliding inside the detector resemble pancake-shaped collections of partons, flying at each other face-on. When one proton interacts with another one, it's actually just one of the partons in one proton that interacts with a parton inside the other proton. As a result, it's hard to know exactly how much energy is involved in a collision, because we don't know which of the partons did the interacting.

Conditions inside an LHC experiment can get pretty intense. There are about fourteen hundred bunches of protons in each beam, and a bunch moving in one direction passes inside the detector by one moving in the other direction about 20 million times per second. Each bunch carries more than 100 billion protons, so that's a lot of particles ready to interact. However, even though the bunches are quite small

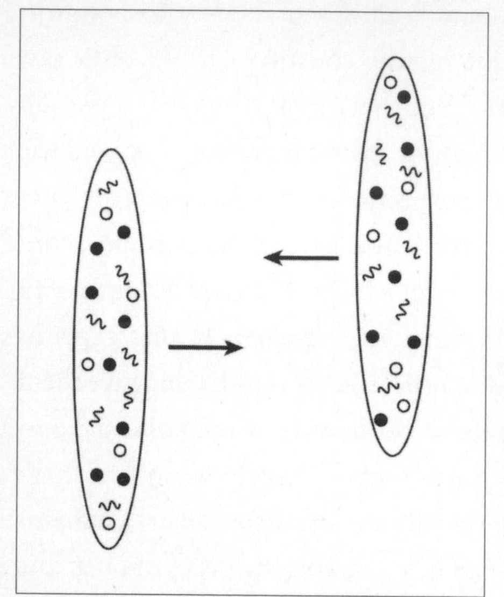

Cartoon of two protons approaching each other in an LHC experiment. The protons, ordinarily spherical, are squeezed into pancake shapes by the effects of relativity, due to moving near the speed of light. Inside the protons are partons, which include quarks (filled circles), antiquarks (empty circles), and gluons (squiggles). There are three more quarks than antiquarks; these are the "valence quarks," while all the other partons are virtual particles.

(a thousandth of an inch across), they are still huge compared with the size of a proton; the overwhelming volume of a bunch is empty space. Each time bunches cross, maybe twenty or so interactions will occur between the billions of protons.

Twenty interactions is still a lot. A single collision of two protons often gives off a messy spray of particles, as many as a hundred hadrons in a single event. We're therefore faced with the danger of "pileup"—many events occurring inside the detector at the same time, making it difficult to distinguish what happened where. This is one of the many reasons why CMS and ATLAS have to strain technology and computing power to the limits of what is currently possible. More collisions are good because it means more data; but too many collisions at once means you can't tell what's going on.

Particles in the chamber

There is a logic to the construction of a particle detector, and it is dictated by the particles themselves. What can possibly come out of a collision? Only the various Standard Model particles we know and love—the six quarks, the six leptons, and the various force-carrying bosons. (We hope to produce completely new species, but those will typically decay into Standard Model particles.) So all we have to do is consider these possibilities and ask how we can best detect and correctly identify them. Let's go through the list.

Quarks

We can lump all the quarks together, because we don't ever see them isolated—they are confined inside hadrons. But you can create a quark-antiquark pair in a collision, and have the two particles move quickly in opposite directions. In that case what happens is that the strong force surrounding the quarks asserts itself, and a spray of hadrons coalesces around the original particle. This shows up in your detector as the "jets" mentioned above. Our job then is to detect the resulting hadrons, which is a relatively easy task, and reconstruct the individual jets, which can be a pain. It can be hard to tell what kind of quark was produced, although there are tricks we can use. For example, bottom quarks last just long enough that they travel a tiny distance before decaying. The particles resulting from the decay therefore emerge slightly offset from the main collision, which can be used to identify bottoms even if their own tracks aren't directly seen.

Gluons

Although they are bosons rather than fermions, gluons are still strongly interacting, so they show up in your detector in a similar way: as a jet of hadrons. One difference is that it's possible to make a single gluon—a quark could spit one off, for example—while newly produced quarks always come paired with antiquarks. So if you see three jets in an event, it means you've made a quark-antiquark pair and a gluon. Events like that are what Sau Lan Wu and her collaborators used to first establish that gluons are real.

W bosons, Z bosons, tau leptons, Higgs bosons

These quite different particles are all grouped together for a simple reason: They are very heavy and therefore short-lived, decaying quickly into other particles—so quickly that they will never be seen directly in your detector. You have to infer their existence by looking at what they decayed into. From this list, tau leptons have the longest lifetime, and in the right circumstances they can last just long enough to be identified.

Electrons and photons

These are the easiest particles to detect and precisely measure. They don't fragment into messy jets like quarks and gluons, but they readily interact with charged particles in a material, creating electrical currents that are simple to identify. It's also straightforward to tell the two apart, because electrons (and positrons, their antiparticles) are electrically charged and therefore swayed by a magnetic field, while photons are neutral and move unimpeded in a straight line.

Neutrinos and gravitons

These are the particles that don't feel either the strong force or the electromagnetic force. As a result, there's no practical way to capture them in a detector, and they just escape unnoticed. Gravitons are only produced by the gravitational interaction, which is so weak that essentially no gravitons are made in a collider and we don't have to worry about them. (In some exotic theories, gravity is effectively strong at high energies and gravitons are produced; physicists certainly take this possibility into account.) Neutrinos, however, are produced by the weak interactions, so they occur all the time. Fortunately they are the only Standard Model particles that can be produced but not detected. So there is a simple rule: Everything that is *not* detected is probably a neutrino.

When two protons collide, they are both moving along the beam pipe, so the total momentum in directions perpendicular to the beam adds up to zero. (The momentum of one particle is the amount of oomph it carries along its direction of motion. For several particles, we just add the separate momenta, but they can combine to zero when the particles are moving in opposite directions.) Momentum is conserved, so it should also add up to zero after the collision. Therefore, we can measure the actual momentum of the particles we detect, and if the answer is not zero, we know there must be neutrinos moving the other way to compensate. This is known as the method of "missing transverse momentum," or just "missing energy." We might not know how many neutrinos carried off the missing momentum, but that can often be deduced from knowing what other particles were produced. (A weak interaction that produces a muon will also make a muon neutrino, and so on.)

Muons

This leaves the muon, which is one of the most intriguing particles from the perspective of an LHC experiment. Like electrons, they leave an easily detectable electrical track, and curve within a magnetic field. But they are two hundred times heavier than an electron. That means they can decay into lighter particles, but their lifetime is still pretty long; unlike the even heavier tau lepton, muons will generally last long enough to survive to the edge of your detector. And they will make it, because muons tend to bash through materials rather than be captured. That's the benefit of being much heavier than electrons but not strongly interacting.

A muon will lumber through all the layers of the experiment like a Jeep driving through a field of wheat, leaving an easily identifiable trail in its wake.

Muons act like super X-rays, penetrating deeply through ordinary stuff. This property was put to good use years ago by Luis Alvarez, who won the Nobel Prize for finding all those hadrons at the Bevatron. Alvarez was intrigued by the pyramids of Egypt, and in particular the large pyramids of the pharaoh Khufu ("Cheops" in Greek) and his son Khafre ("Chephren"), which sit next to each other in Giza, outside Cairo. Khufu's is the "Great Pyramid," and it was originally slightly larger, although external wear has left it a bit smaller than Khafre's these days. Inside Khufu's pyramid are three chambers, while Khafre's pyramid seems solid save for one burial chamber at ground level. This difference has puzzled archaeologists for years, and many have theorized that Khafre's pyramid contains undiscovered chambers.

Alvarez, a brilliant physicist with a penchant for puzzles, hit on an idea: Use muons coming from the skies in the form of cosmic rays to peek inside the rock of Khafre's pyramid. It would be a crude experiment, but able to distinguish between solid rock and an empty chamber. Alvarez's team of Egyptian and American physicists assembled a muon detector in the single known chamber at the lower level of the pyramid, looking to count the number of muons coming from different directions. This was 1967, and the project suffered a delay when the Arab-Israeli war broke out the day before they were scheduled to first take data. But eventually they got up and running, and discovered . . . nothing. All directions of the pyramid appeared to be equally good at stopping muons, in contrast to the hope that some directions would let more of them through because they contained an empty chamber. It remains a puzzle why the son's pyramid is noticeably less complicated than the father's.

Layers of detectors

The ATLAS and CMS experiments settled on a strategy for squeezing as much information as possible out of the particle collisions they observe. Both detectors are constructed in layers, with four different pieces of apparatus serving very specific purposes: an inner detector, surrounded by an electromagnetic calorimeter, which is in turn surrounded by a hadronic calorimeter, and finally a muon detector on the outside. Any particles produced in a collision will radiate outward from the collision point, passing through different layers until they are finally captured or they escape to the external world.

The job of the inner detector, the innermost layer of the onion, is to act as a tracker that provides pinpoint information about the trajectories of charged particles that emerge from the collision point. It's

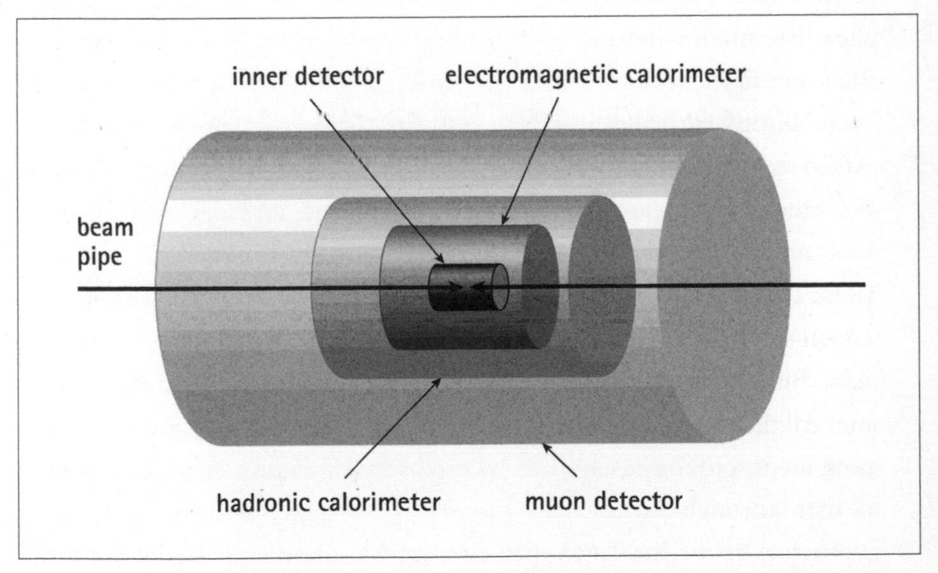

A cartoon depiction of a general-purpose particle experiment, such as ATLAS or CMS. The central region contains an inner detector that measures the paths of charged particles. Next is the electromagnetic calorimeter that captures photons and electrons, then the hadronic calorimeter that captures hadrons. Finally, the muon detector that tracks the muons.

not an easy job; every square centimeter of the instrument is bombarded with tens of millions of particles per second. Anything you put there has to do its job while surviving an unheard-of amount of radiation exposure. Indeed, the very first design drawings for CMS simply left this region of their detector empty, since physicists didn't think they could build a precision instrument that could take the heat. Fortunately, they were encouraged to keep trying by rumors that the military had solved the problem of making electronic readouts that could function effectively in this kind of harsh environment. They ultimately succeeded by figuring out how to "harden" very fine commercial electronics that weren't originally intended to withstand such conditions.

The inner detectors are complicated multicomponent machines with slightly different features between the two experiments. The ATLAS inner detector, for example, consists of three different instruments: a pixel detector with incredibly fine resolution; a semiconductor tracker made of silicon strips; and a transition radiation tracker made of gold-plated tungsten wire inside thin tubes known as "straws." The job of the inner detector is to record the paths of emerging particles as precisely as possible, allowing physicists to reconstruct the interaction points from which they originate.

The next layers are the calorimeters, electronic and hadronic. "Calorimeter" is a fancy word for "device that measures energy," just as "calories" are used to quantify the energy in the foods we eat. The electromagnetic calorimeter is able to capture electrons and photons via their interactions with nuclei and electrons in the calorimeter itself. Strongly interacting particles generally pass right through the electromagnetic calorimeter, only to be captured by the hadronic calorimeter. This component consists of layers of dense metal that interact with the hadrons, alternating with scintillators that measure the amount of energy deposited. Measuring the energies of the particles is a crucial step in identifying what they are, and often the mass of whatever particle decayed to create them.

The final layers of the experiments are the muon detectors. Muons

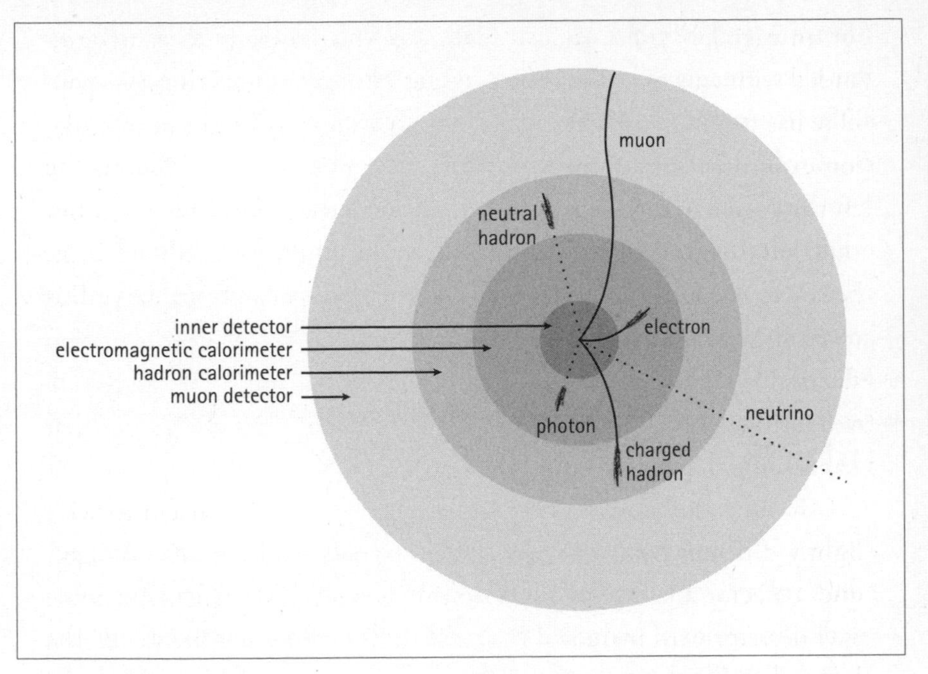

A cross-section of an experiment, showing the behavior of different particles. Neutral particles like photons and neutral hadrons are invisible to the inner detector, but charged particles leave curved tracks. Photons and electrons are captured by the electromagnetic calorimeter, while hadrons are captured by the hadron calorimeter. Muons make it to the outer detector, and neutrinos escape detection entirely. In the CMS experiment, muons curve in the opposite direction in the outer detector because the magnetic field points the opposite way.

have enough momentum to punch through the calorimeters, but can be precisely measured by the giant magnetic chambers that surround them. This is important because muons are not created by the strong interactions (since they are leptons, not quarks), and only rarely by the electromagnetic interactions (because they are so heavy and it's easier just to make electrons). Therefore, muons generally come about from the weak interactions, or something brand-new. Either alternative is interesting, and muons play an important role in the search for the Higgs.

We now see why the design of the ATLAS and CMS experiments takes the form that it does. The inner detectors provide precision information about the trajectories of all charged particles leaving the collisions.

Electrons and photons are captured, and their energies measured, by the electromagnetic calorimeter, while strongly interacting particles suffer the same fate in the hadronic calorimeter. Muons escape the calorimeters but are carefully studied in the muon detector. Among the known particles, only neutrinos escape undetected, and we can infer their existence by looking for missing momentum. All in all, an ingenious scheme to squeeze out all the information we can from the colliding protons produced by the LHC.

Information overload

At the LHC, bunches of protons come into collision 20 million times per second. Every crossing produces dozens of collisions, so we have hundreds of millions of collisions a second. Every collision is like fireworks going off inside the detector, creating multiple particles, up to a hundred or more. And the finely calibrated instruments inside the experiments collect precise information about what every one of those particles does.

That's a lot of information. A single collision event at the LHC results in about one megabyte of data. (The raw data is more than twenty megabytes, but clever compression brings it close to a single megabyte.) That's the size of the text of a large book, or the total amount of RAM in a space shuttle's operating system. Decent home-computer hard drives these days can store a terabyte of data, or a million megabytes, which is huge—all the text in all the books of the Library of Congress amounts to only about twenty terabytes. You could store a million LHC events on one of these ordinary hard drives, which sounds good—except when you remember that there are hundreds of millions of events per second. So you would be filling up a thousand hard drives per second. Not really feasible, even given that CERN can afford better hard drives than you have on your laptop.

Outside the LHC, the largest single database in the world belongs to the World Data Center for Climate in Germany. It contains about six petabytes of climate data, or six thousand terabytes. If we recorded all the data created at the LHC, we would overflow a database of that size in a couple of seconds. Welcome to the world of Big Data.

Clearly, data storage (and transmission and analysis) at the LHC is a major challenge, one that is met by a combination of many different techniques. The most important one, however, is the most basic: not recording the data in the first place. This is worth emphasizing: *The overwhelming majority of data collected by the LHC is instantly thrown away*. We have no choice; there is no feasible way to record it all.

You might think that a more cost-effective strategy would simply be to not produce so much data in the first place, for example by lowering the luminosity of the machine. But particle physics doesn't work that way—every collision is important, even if we don't record its data to disk. That's because quantum mechanics, which is ultimately responsible for the interactions that create these particles, only predicts the probability of certain outcomes. We can't pick and choose what comes out when we collide two protons; we have to take what nature gives us. A large majority of the time, what nature gives us is pretty boring, at least in the sense that it's stuff we already understand. To create a small number of interesting events, we have to produce an enormous number of pedestrian events, and swiftly pick out the interesting nuggets.

This raises a different problem, of course: how to figure out whether an event is "interesting," and to do so extremely quickly, so that we can decide whether this is data worth keeping. That's the job of the trigger, one of the most crucial aspects of an LHC experiment.

The trigger itself is a combination of hardware and software. The first-level trigger brings the output of all the instruments in the experiment into an electronic buffer and performs an ultrarapid scan (in about a microsecond) to see if anything potentially interesting is going on. About ten thousand events out of a billion get a stamp of approval and move on. The second-level trigger is a sophisticated piece of software that looks at more precise characterizations of the events (much like an ER doctor making a preliminary rapid diagnosis, then homing in with more delicate tests) to get you down to the events that are actually recorded for later analysis. We end up keeping only several hundred events out of the many millions produced per second—but they're the most interesting ones.

As you might guess, a lot of hard thought and spirited disagreement go into deciding which events to keep and which to toss. It's natural to worry that some real gems are being thrown away in all that discarded data, so the physicists at CMS and ATLAS are constantly working to refine their triggers in response to both improved experimental know-how and novel ideas from the theorists.

Sharing data

Even after running everything through the trigger, we're still left with a hundred events per second, each characterized by about a megabyte of data. Now we have to analyze it. And by "we," I mean "the thousands of members of the ATLAS and CMS experiments, working at institutions all over the world" (which don't actually include me). For the physicists to analyze the data, they need access to it, which means a challenge in information transmission. Fortunately, this issue was anticipated for years, and physicists and computer scientists have worked hard to construct a Worldwide LHC Computing Grid that connects computing centers in thirty-five different countries, using a combination of the public Internet and private optical cables. In 2003, a new land speed record for data was set when more than a terabyte of information traveled more than five thousand miles from CERN to Caltech in under thirty minutes. That's like downloading a full-length feature film in seven seconds.

This kind of crazy speed is necessary; in 2010, the four main experiments at the LHC produced more than thirteen petabytes of data.

The Grid, as it is affectionately known, takes this data and parcels it out to different computing centers around the world, arranged in a series of tiers. Tier 0 is CERN itself. There are eleven Tier 1 sites, which play an important role in sifting through and classifying the data, and 140 Tier 2 sites, where specific analysis tasks are performed. This way every physicist in the world who wants to analyze LHC data doesn't have to connect directly to CERN, running the risk of breaking the Internet for good.

Necessity is the mother of invention. It should come as no surprise that the unique data challenges presented by particle physics have led to unique solutions. One of those solutions, from many years ago now, has changed the way we all live: the World Wide Web. The Web originated in a 1989 proposal from Tim Berners-Lee, who at the time was working at CERN, and is currently director of the World Wide Web Consortium. Berners-Lee thought it would be useful for physicists at the lab to have access to different kinds of information, stored on distributed computers, through a hypertext system based on Web documents and links between them. The WWW is this system of interlinked files, built on top of the data-sharing network we call the Internet (for which we can't give CERN any credit). The Web as we currently know it, and all the effects it has had on our lives, are spinoffs from basic research in particle physics.

Fabiola Gianotti, the Italian physicist who is the current leader of ATLAS, told me that the most pleasant surprise when the LHC first turned on wasn't the performance of her experiment, although that was quite impressive—it was that the data transmission system functioned flawlessly right from the start. Not that the process has been entirely without challenges. In September 2008, soon after the first particles had circulated in the LHC, the computing system at CMS was hacked by a group billing itself as the "Greek Security Team." They did no real damage, and in fact claimed to be performing a public service, as they replaced a Web page with a warning in Greek that said, "We're pulling your pants down because we don't want to see you running around naked looking to hide yourselves when the panic comes." Order was quickly restored, and the disturbance didn't delay the experiment in any way—although it maybe prompted a closer look at Internet security throughout CERN.

With the LHC itself humming along, CMS and ATLAS running at the peak of their capabilities, and data being rapidly shared and analyzed around the globe, all the pieces are in place for a full-on assault on the important questions in particle physics. One new particle is in the bag, and we're looking for more.

SEVEN
PARTICLES IN THE WAVES

In which we suggest that everything in the universe is made out of fields: force fields that push and pull, and matter fields whose vibrations are particles.

The Insane Clown Posse, a hip-hop duo known for their provocative lyrics and scary clown makeup, caused a stir in 2010 with their single "Miracles." At this point in their career, Violent J and Shaggy 2 Dope (not the names they were born with) were no strangers to controversy. They had engaged in a feud with Eminem, explored an unsuccessful stint as professional wrestlers, and once gave a brief concert to a bewildered audience only to find out that they were in the wrong building. Their songs tell stories of necrophilia and cannibalism, and in one case said mean things about Santa Claus. Also, Violent J was arrested after a show for hitting an audience member thirty times with his microphone.

But the "Miracles" controversy was something different. The lads weren't aiming to shock but to share their wonder at the world around us. It came out like this:

Stop and look around, it's all astounding
Water, fire, air and dirt

*F***ing magnets,*
How do they work?

Through the magic of the Internet, this little snippet gained quite a bit of notoriety, especially from scientifically minded types who were eager to point out that we actually have a pretty good idea of how magnets work.

I would like to stand up just a tiny bit for Insane Clown Posse. Yes, we've understood magnetism for quite some time, and scientific investigation generally enhances our appreciation of natural phenomena rather than draining the magic out of everything. However, they have put their fingers on an important fact we may be too quick to overlook: Magnets are actually pretty astounding.

What's amazing about magnets is not that they stick to metal—lots of things stick to lots of other things, from geckos to pieces of chewing gum. What's amazing is that, when you bring a magnet close to a piece of metal, you can feel it being attracted *before* they're actually touching. Magnets aren't like adhesive tape or glue, which must be in contact with something before sticking. Magnets reach out, across apparently empty space, to pull things toward them. Kind of freaky, when you think about it.

Physicists call this type of thing "action at a distance," and it used to bother the world's greatest minds as much as it bothers Violent J and Shaggy 2 Dope. These days we are less bothered, because we've figured out that the space across which the magnet is apparently reaching isn't really "empty" at all. It's filled with a magnetic field—invisible lines of force that reach out from the magnet—ready to grab ahold of any susceptible object that might come their way. We can make these lines of force seem more tangible by putting the magnet in the presence of some small iron filings, which line up with the magnetic field in beautiful patterns.

The important point is that the magnetic field is there whether or not it's grabbing on to anything. If there is a magnet, there is a magnetic field that surrounds it, even though we can't see it. The field is strong when we're close to the magnet, and weaker far away. In fact, there's a magnetic field at absolutely every point in space, regardless of whether there are any magnets nearby. The field might be quite small—or even precisely zero—but at every point there is some answer to the question "What's the value of the magnetic field here?" (It really is "the" magnetic field, not a separate one for every magnet; put two magnets near each other and their fields just add together.)

I'm not sure if the Insane Clown Posse wants to hear it, but the importance of fields extends well beyond magnets. The world is really made out of fields. Sometimes the stuff of the universe looks like particles, due to the peculiarities of quantum mechanics, but deep down it's really fields. Empty space isn't as empty as it looks. At every point there is a rich collection of fields, each taking on some value or another—or more precisely, due to the uncertainty that accompanies quantum mechanics, a distribution of possible values we could potentially observe.

When we talk about particle physics, we don't usually emphasize that we're actually talking about field physics. But we are. The point of this chapter is to reorient our intuition, in order to appreciate how quantum fields are the ultimate building blocks of reality as we currently understand it.

The fields themselves aren't "made of" anything—fields are what the world is made of. We don't know of any lower level of reality. (Maybe string theory, but that's still hypothetical.) Magnetism is carried by a field, as are gravity and the nuclear forces. Even what we call "matter"—particles like electrons and protons—is really just a set of vibrating fields. The particle we call the "Higgs boson" is important, but not so much for its own sake; what matters is the Higgs field from which it springs, which plays a central role in how our universe works. Astounding indeed.

In the first few chapters we gave a brief introduction to the particles of the Standard Model, and mentioned that all particles arise as vibrations in fields. In the past few chapters we looked at the accelerators and detectors that help us explore the subatomic world, including the LHC. In this chapter and the next we're going to back up a bit, taking a closer look at the idea of a field, how particles arise from fields, how symmetries give rise to forces, and how the Higgs field can break a symmetry and give us the variety of particles we see. That will put us in perfect position to understand how experimentalists hunt for the Higgs, and what it means that we've found it.

The gravitational field

These days we recognize that fields are all around us, but it took a while for scientists to start thinking in terms of "field theory." You might guess that the idea of a gravitational field is even more obvious than the idea of a magnetic field, and you'd be right. But it's not *completely* obvious.

The most famous story about gravity involves Isaac Newton and an apple that supposedly fell on his head, inspiring him to concoct his theory of universal gravitation. (It's mostly famous because Newton himself couldn't stop telling it later in life, in an unnecessary attempt to add some extra juice to his reputation as a genius.) The simplest version of the anecdote says that the apple helped Newton "invent" or perhaps "discover" gravity, although a moment's contemplation reveals that this doesn't quite make sense. People knew about gravity before Newton came along—it's not like nobody had noticed that apples fall down, not up.

What came to Newton was the connection between the fall of the apple and the motion of the planets. He didn't invent gravity, but he realized that it was *universal*—the gravitational attraction that keeps the planets orbiting around the sun and the moon orbiting around the earth was the same force that pulls apples toward the ground. You might not think that even this is the kind of insight of which legends

JoAnne Hewett, rapping about dark matter at a physics slam in Eugene, Oregon, in 2011.

At CERN on July 4, 2012, Fabiola Gianotti, Rolf Heuer, and Joe Incandela, preparing for the big announcement.

Leon Lederman, standing outside Fermilab.

FERMILAB NATIONAL ACCELERATOR LABORATORY

Sau Lan Wu of the University of Wisconsin, who has been searching for the Higgs at both LEP and the LHC.

JEFF MILLER/UNIVERSITY OF WISCONSIN

Carlo Rubbia, discoverer of the W and Z bosons and advocate for the LHC.

CREATIVE COMMONS

Fabiola Gianotti, spokesperson for ATLAS in 2011 and 2012.

© CERN

Lyn Evans, the man who built the LHC.

An aerial view of CERN and the Large Hadron Collider, with major experiments marked. The actual ring is underground and not visible from above.

The Globe of Science and Innovation at CERN, a striking building that serves as a symbol for the laboratory. The Globe houses public-information exhibitions about particle physics and CERN's mission.

Inside the LHC tunnel, with dipole magnets installed and ready to go.

Damage to LHC magnets after the September 19 accident.

All the protons for the LHC beam come from this tiny canister of hydrogen. It contains enough protons to feed the LHC for a billion years.

© CERN

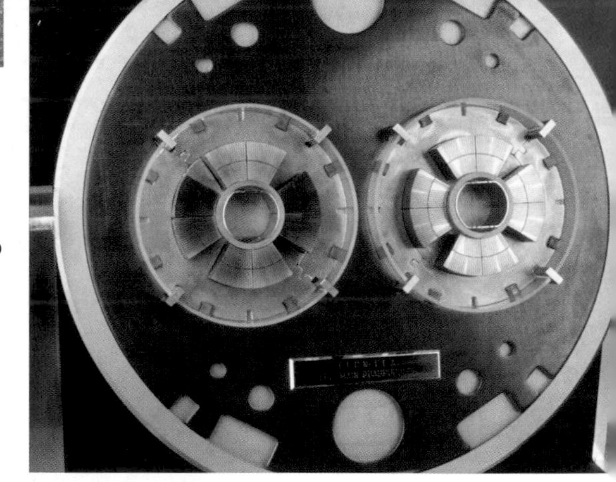

A model of the cross-section through the dipole magnets in the LHC. The two beam pipes carry protons moving in opposite directions.

© CERN

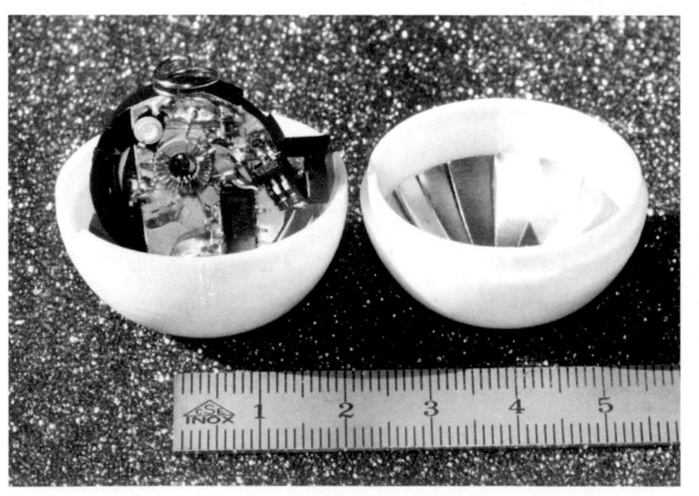

One of the "Ping-Pong balls" equipped with a radio transmitter that was sent down the beam pipe of the LHC to check for obstructions.

LYN EVANS

are made. After all, something keeps the planets from zooming off into interstellar space, and something pulls apples to the ground, so why shouldn't they be the same thing?

If that's what you're thinking, it's only because you live in a post-Newtonian world. Before Newton came along, we wouldn't have blamed the earth's pull for the fall of the apple—we would have blamed the apple itself. Aristotle, for example, thought that different kinds of matter all had natural states of being. The natural state of a massive body was to be on the ground. If it is lifted above the ground, it wants to fall.

This idea that falling is due to an object's natural inclination rather than the earth pulling on it is actually quite intuitive. I once served as a science consultant on a big-budget Hollywood movie, for which the designers thought it would be cool to portray a thrilling fight scene on a planet that was shaped like a disk, rather than a sphere. And it would be cool, you can't argue with that. But they planned to have the scene climax with the bad guys falling off the edge of the planet. Pulled by . . . what, exactly? If you think of falling as something that things naturally do, rather than as a consequence of some large object pulling on them due to gravity, it's a natural mistake to make. (But we managed to keep it out of the movie.)

Newton suggested that every object in the universe exerts a gravitational pull on every other object in the universe. Heavier objects exert a greater pull, and nearby objects are pulled more strongly than faraway ones. This idea fits the data beautifully, and represents a marvelous unification of what happens on earth and what happens in the sky.

But Newton's theory of gravity bugged a lot of people. How does the moon, for example, *know* that the earth is exerting a gravitational pull on it? Earth is very far away, after all, and we're used to forces being exerted when we bump into things, not when we're elsewhere in the universe. This is the puzzle of "action at a distance," and it disturbed Newton as well as his critics. At some point, however, when your theory does an amazingly good job at explaining a multitude of phenomena, you shrug your shoulders and admit that nature apparently just works that way. It's pretty much the situation we're in with quantum mechanics today: a theory that fits the data, but which we don't think we understand as well as we should.

It wasn't until the late 1700s that a French physicist, Pierre-Simon Laplace, showed that you didn't have to think of Newtonian gravity in terms of magical action at a distance. Laplace realized that you could imagine a field filling all of space, later dubbed the "gravitational potential field." The gravitational potential is distorted by massive bodies, just like the temperature of the air in a room is affected by a hot oven; the distortion is strong nearby, and fades as we get farther away. The force due to gravity arises because objects are pushed by the field itself: They feel a tug toward the direction in which the gravitational potential field is decreasing, much like a ball placed on an uneven surface will start rolling the direction in which the height of the surface is decreasing.

Mathematically, Laplace's theory is identical to Newton's. But conceptually, it fits in much better with our intuition that all physics, like politics, is local. It's not that earth just reaches out and attracts the moon; earth affects the gravitational potential nearby, and that affects the potential right next door, and onward in a smooth sequence all the way to the moon (and beyond). The force of gravity isn't a mysterious effect that leaps over infinite distances; it arises from the smooth variation of an invisible field that permeates all of space.

The electromagnetic field

It was in the study of electromagnetism where the idea of fields came into its own. There is an electric field, and also a magnetic field, but physicists just say "electromagnetism" as a single word to indicate that they are really two different manifestations of a single underlying field. The connection between the two wasn't always so obvious.

Magnetism had been known since ancient times, of course; the Han dynasty in China had developed magnetic compasses more than two thousand years ago. And electricity had been recognized, both in the form of shocks you could receive from eels and the static electricity that collects on amber when it is rubbed with a cloth. There were even some hints that the phenomena were related; Benjamin Franklin, in between flying kites and fomenting rebellions, showed that it was possible to magnetize needles with electricity.

But the ideas didn't truly come together until 1820, when a Danish physicist named Hans Christian Ørsted was giving a lecture on the nature of electricity and magnetism. Ørsted had thought of a clever way to demonstrate the hypothetical connection between the two: He would build an electrical circuit, and then run the current next to a magnet and see if its needle was deflected from true north by the running electricity. Unfortunately, an accident prevented Ørsted from actually carrying out the experiment before it was time for his lecture. He decided to simply do the experiment right there in front of the assembled crowd, convinced that it must work . . . and it did. He flipped a switch, electrical current flowed through a wire, and he saw a small but unmistakable jitter in the compass needle. According the Ørsted's own account, the effect was quite small, and the audience went away unimpressed. But from that day forward, electricity and magnetism had merged into the subject of electromagnetism.

Through subsequent work by people such as Michael Faraday and James Clerk Maxwell, a sophisticated theory of the electromagnetic field was developed. Once this theory was in place, we could answer questions about the dynamics of that field. For example, what happens when you take an electric charge and shake it up and down? (The same question could have been asked about gravity, but the gravitational force is so weak it would be very hard to answer the question experimentally.)

What happens when you shake a charge is, quite naturally, that you create ripples in the electromagnetic field. And these ripples propagate outward as waves, much like waves on water when you drop a stone into it. There is a good name for these electromagnetic waves: light. When we turn on a light switch, what happens is that electrical current flows through the filament of the lightbulb, heating it up. That heating shakes up the atoms in the filament and their associated electrons, causing them to jiggle back and forth. That jiggling sets up waves in the electromagnetic field that travel to our eyes and are perceived as light.

The identification of light with waves in the electromagnetic field represents another great triumph of unification in physics. It went further when we realized that what we call visible light is only particular wavelengths of radiation—those that can be seen by the human eye. Shorter wavelengths include X-rays and ultraviolet light, while longer wavelengths include infrared light, microwaves, and radio waves. The work of Faraday and Maxwell received spectacular confirmation in 1888, when the German physicist Heinrich Hertz was able to produce and detect radio waves for the first time.

When you use a remote control to turn on your TV, it looks like action at a distance, but it's really not. You push the button and an electrical current starts to jiggle back and forth inside a circuit in the remote, creating a radio wave that propagates through the electromagnetic field to the TV and is absorbed by a similar gizmo. In the modern world, the electromagnetic field around us is made to do an enormous amount of work—illuminating our environment, sending signals to our cell phones and wireless computers, and microwaving our food. In every case it's set up by moving charges that send ripples out through the field. All of which, by the way, was completely unanticipated by Hertz. When he was asked what his radio-wave-detecting device would ultimately be good for, he replied, "It's of no use whatsoever." Prodded to suggest some practical application, he replied, "Nothing, I guess." Something to keep in mind as we contemplate the eventual applications of basic research.

Waves of gravity

It wasn't until after physicists understood the relationship between electromagnetism and light that they began to wonder whether a similar phenomenon should happen with gravity. It might seem like an academic question, since you need an object the size of a planet or moon to create a gravitational field big enough to measure. We're not going to pick up the earth and shake it back and forth to make waves. But to the universe, that's no problem at all. Our galaxy is full of binary stars—systems where two stars orbit around each other—presumably shaking the gravitational field as they go. Does that lead to rippling waves spreading in every direction?

Interestingly, gravity as Newton or Laplace described it would *not* predict radiation of any kind. When a planet or star moves, the theory says that its gravitational pull changes instantaneously all across the universe. It's not a propagating wave but an instant transformation everywhere.

That's just one way in which Newtonian gravity doesn't seem to fit well with the changing framework of physics that developed over the course of the nineteenth century. Electromagnetism, and especially the central role played by the speed of light, was instrumental in inspiring Albert Einstein and others to develop the theory of special relativity in 1905. According to that theory, nothing can travel faster than light—not even hypothetical changes in the gravitational field. Something would have to give. After ten years of hard work, Einstein was able to construct a brand-new theory of gravity, known as "general relativity," that replaced Newton's entirely.

Just like Laplace's version of Newtonian gravity, Einstein's general relativity describes gravity in terms of a field that is defined at every point in space. But Einstein's field is a much more mathematically complicated and intimidating field than Laplace's; rather than the gravitational potential, which is just a single number at each point, Einstein used something called the "metric tensor," which can be thought of as a collection of ten independent numbers at every point. This mathematical complexity helps general relativity accrue its reputation as a very difficult theory to understand. But the basic idea is simple, if profound: The metric describes the curvature of spacetime itself. According to Einstein, gravity is a manifestation of the bending and stretching of the very fabric of space, the way we measure distances and times in the universe. When we say, "The gravitational field is zero," we mean that spacetime is flat, and the Euclidean geometry we learned in high school is valid.

One happy consequence of general relativity is that, just like with electromagnetism, ripples in the field describe waves traveling at the speed of light. And we have detected them, although not directly. In 1974, Russell Hulse and Joseph Taylor discovered a binary system in which both objects are neutron stars, rapidly spinning in a very tight orbit. General relativity predicts that such a system should lose energy by giving off gravitational waves, causing the orbital period to gradually decrease as the stars draw closer together. Hulse and Taylor were able to measure this change in the period, exactly as predicted by Einstein's theory; in 1993, they were awarded the Nobel Prize in Physics for their efforts.

That's an indirect measurement of gravitational waves, rather than directly seeing their effects in a laboratory here on earth. We are certainly trying. There are a number of ongoing efforts to observe gravitational waves coming from astrophysical sources, typically by bouncing lasers off mirrors separated by several kilometers. As a gravitational wave passes by, it stretches spacetime, bringing the mirrors closer together and then farther apart. That can be detected by measuring tiny changes in the number of laser wavelengths between the two mirrors. In the United States, the Laser Interferometer Gravitational Wave Observatory (LIGO) consists of two separate facilities, one in Washington State and the other in Louisiana. They collaborate with the VIRGO observatory in Italy and GEO600 in Germany. None of these laboratories has yet detected gravitational waves—but scientists are very optimistic that recent upgrades will help them make a dramatic discovery. If and when they do, it will be vivid confirmation that gravity is communicated by a dynamic, vibrating field.

Particles out of fields

The realization that light is an electromagnetic wave flew in the face of Newton's theory of light, which insisted that it was made of particles dubbed "corpuscles." There were good arguments on both sides. On the one hand, light casts a sharp shadow, like you might expect from a spray of particles, rather than bending around corners, as our experience with water and sound waves might lead us to believe. On the other hand, light can form interference patterns when passing through narrow openings, as a wave would do. The electromagnetic synthesis seemed to clinch the issue in favor of waves.

Conceptually, a field is the opposite of a particle. A particle has a specific location in space, while a field exists at every point in space. It's defined by its magnitude, which is some particular number at every point, and maybe some other qualities like a direction. Quantum mechanics, which was born in 1900 and came to dominate the physics of the twentieth century, ultimately brought the two concepts together. Long story short: Everything is made of fields, but when we look at them closely we see particles.

Imagine you are outside on a very dark night, watching a friend holding a candle walk away from you. The candle grows dimmer as the distance to your friend increases. Eventually it becomes so dim that you can't see it at all. But, you might think, that's due to the fact that our eyes are imperfect instruments. Perhaps if we had ideal vision, we would see the candle grow progressively dimmer but never quite go away entirely.

Actually that's not what would happen. With perfect eyes, we would see the candle grow dimmer for a while, but at some point a remarkable thing would happen. Rather than growing progressively more faint, the candlelight would begin to flicker on and off, with some fixed brightness while it was on. As your friend retreated, the off periods would lengthen with respect to the on periods; eventually the candle would be almost completely dark, save for very rare flashes of low-intensity light. Those flashes would be individual particles of light: photons. Physicist David Deutsch discusses this thought experiment in his book *The Fabric of Reality*, where he notes that frogs have better vision than humans, good enough to see individual photons.

The idea behind photons stretches back to Max Planck and Albert Einstein at the turn of the last century. Planck was thinking about the radiation that objects give off when they are heated. The wave theory of light predicted there should be much more radiation coming out with very short wavelengths, and therefore very high energies, than we actually observe. Planck suggested a brilliant and somewhat startling way out: that light came in discrete packets, or quanta (plural of "quantum"), and that a light quantum with some fixed wavelength would have a fixed energy. You need a good amount of energy to make just one quantum of short-wavelength light; Planck's idea therefore helped explain why there is so much less radiation at short wavelengths than the wave theory predicted.

This connection between energy and wavelength is a key concept in quantum mechanics and field theory. The wavelength is just the distance between two successive crests of a wave. When it's short, the wave is all bunched together. It costs energy to do that, so we see why Planck's packets of light have high energy when the corresponding wavelength is short, as in ultraviolet light or X-rays. Long wavelengths, like radio waves, imply individual light quanta with very low energies. Once quantum mechanics was invented, this relationship could even be extended to massive particles. High mass implies short wavelength, which means that a particle takes up less space. That's why it's the electrons, not the protons or neutrons, that define the size of an atom; they're the lightest particle involved, so they have the longest wavelength, and therefore take up the most space. In a sense, it's even why the LHC has to be so big. We're trying to look at things that happen within very short distances, which means we need to use very small wavelengths, which means we need highly energetic particles, which means we need a giant accelerator to get them moving as fast as possible.

Planck didn't make the leap from quantized energies to literal particles of light. He thought of his idea as a sort of trick to get the right answer, not as a fundamental part of how reality works. That step was taken by Einstein, who was puzzling over something called the "photoelectric effect." When you shine bright light on metal, you can shake electrons loose from the metal's atoms. You might think that the number of electrons shaken free would depend on the intensity of the light, because more energy comes in when the light beam is more intense. But that's not quite right; when the light has a long wavelength even a bright source doesn't shake loose any electrons at all, while short-wavelength light is able to shake some loose even when it's quite dim. Einstein realized that the photoelectric effect could be explained if we believed that light always came in individual quanta rather than as a smooth wave—not only when it was emitted by glowing bodies. "High intensity but long wavelength" implies a barrage of quanta, but each with an energy that is too small to disturb any electrons at all; "low intensity but short wavelength" means just a few quanta, but each with enough energy to do the job.

Neither Planck nor Einstein used the word "photon." That was coined by Gilbert Lewis in the 1920s, and popularized by Arthur Compton. It was Compton who finally convinced people that light came in the form of particles, by showing that the light quanta had momentum as well as energy.

Einstein's paper on the photoelectric effect was the work for which he ultimately won the Nobel Prize. It was published in 1905, and Einstein had another paper in the very same issue of the journal where it appeared—his other paper was the one that formulated the special theory of relativity. That's what it was like to be Einstein in 1905: You publish a groundbreaking paper that helps lay the foundations of quantum mechanics, and for which you later win the Nobel Prize, but it's only the second-most important paper that you publish in that issue of the journal.

Quantum implications

Quantum mechanics sneaked up on physicists over the course of the early decades of the twentieth century. Starting with Planck and Einstein, people tried to make sense of the behavior of photons and atoms, and by the time they were done they had completely upended the reliable Newtonian view of the world. There have been many revolutions in physics, but two stand out far above the rest: when Newton put together his great vision of "classical" mechanics in the 1600s, and when a collection of brilliant scientists worked together to replace Newton's theory with that of quantum mechanics.

The major difference between the quantum world and the classical one lies in the relationship between what "really exists" and what we can actually observe. Of course any real-world measurement is subject to the imprecision of our measuring devices, but in classical mechanics we can at least imagine being more and more careful and bringing our measurements closer and closer to reality. Quantum mechanics denies us that possibility, even in principle. In the quantum world, what we can possibly see is only a small subset of what really exists.

Here is a ham-fisted analogy to illustrate the point. Imagine you have a friend who is very photogenic, but you notice something unusual about pictures in which she appears—she is always precisely in profile, showing her left side or right side but never appearing from the front or back. When you see her from the side and then take a picture, the image is always correctly from that side. But when you see her from directly in front and then take a picture, half the time it comes out as her left profile and half the time as her right profile. (The terms of the analogy dictate that "taking a picture" is equivalent to "making a quantum observation.") You can take a picture from one angle and then really quickly move around to take a picture from ninety degrees away—but you only ever capture her in profile. That's the essence of quantum mechanics—our friend can really be in any orientation, but when we snap a photo we see only one of two possible angles. This is a good analogy for the "spin" of an electron in quantum mechanics, a property we only ever measure to be precisely clockwise or counterclockwise, no matter what axis we use to make the measurement.

The same principle holds for other observable quantities. Consider the location of a particle. In classical mechanics, there is something called the "particle's position," and we can measure it. In quantum mechanics there is no such thing. Instead, there is something called the "wave function" of the particle, which is a set of numbers that reveal the *probability* of seeing the particle in any particular place when we look at it. There is no such thing as "where the particle is, really"—but when we look, we always see it in some particular place.

When quantum mechanics gets applied to fields, we end up with "quantum field theory," which is the basis for our modern explanations of reality at its most fundamental level. According to quantum field theory, when we observe a field carefully enough we see it resolve into individual particles—although the field itself is real. (The field actually has a wave function describing the probability of finding it with any particular value at each point in space.) Think of a TV set or computer monitor, which seems to display a smooth picture from a distance, but close up we find that it's actually a collection of tiny pixels. On a quantum TV set there really is a smooth picture, but when we look closely at it we can only ever observe it as pixels.

Quantum field theory is responsible for the phenomenon of virtual particles, including the partons (quarks and gluons) inside protons that are so crucial to what happens in LHC collisions. Just as we can never quite pin down a single particle to a definite position, we can never really pin a field down to a definite configuration. If we look at it closely enough, we see particles appearing and disappearing in empty space, depending on the local conditions. Virtual particles are a direct consequence of the uncertainty inherent in quantum measurement.

Physics students for generations now have been confronted with the ominous-sounding question, "Is matter really made of particles or waves?" Often they get through years of education without quite grasping the answer. Here it is: Matter is really waves (quantum fields), but when we look at it carefully enough we see particles. If only our eyes were as sensitive as those of frogs, this might make more sense to us.

Matter from fields

So light is a wave, a set of propagating ripples in the electromagnetic field that pervades space. When we throw quantum mechanics into the mix, we end up with quantum field theory, which says that when we look closely at an electromagnetic field we see it as individual photons. The same logic works for gravity—it's described by a field, and there are gravitational waves that move through space at the speed of light, and if we looked at such a wave carefully enough we would see it as a collection of massless particles called "gravitons." Gravity is far too weak for us to imagine detecting individual gravitons, but the basic truth of quantum mechanics insists that they must be there. Likewise, the strong nuclear force is carried by a field that we observe as particles called "gluons," and the weak nuclear force is a field carried by W and Z bosons.

All well and good; once we get that forces arise from fields stretching through space, and that quantum mechanics makes fields look like particles, we have a pretty good grasp of how the forces of nature work. But what about the matter that those forces act upon? It's one thing to think of gravity or magnetism as arising from a field, but something else entirely to think of atoms themselves as being associated with fields. If anything is truly a particle rather than a field, it's one of those tiny electrons that orbits around atoms. Right?

Wrong. Just like force-carrying particles, matter particles also arise from applying the rules of quantum mechanics to a field that fills space. As we've discussed, force-carrying particles are bosons, while matter particles are fermions. They correspond to different kinds of fields, but fields nevertheless.

Bosons can pile on top of one another, while fermions take up space. Let's think about this from the point of view of the fields of which those particles are vibrations. The difference between them comes down to a simple distinction: Boson fields can take on any value whatsoever, while each possible vibrational frequency of a fermion field is either "on" or "off," once and for all. When a boson field like the electromagnetic field has a really large value, it corresponds to a large number of particles; when it's a small but nonzero value, it's just a few particles. Those possibilities aren't open to fermion fields. There is either a particle there (in some particular state), or there isn't. This crucial feature is known as the "Pauli exclusion principle": No two fermion particles can be in the same state. To define the "state" of a particle we need to tell you where it is, what energy it has, and maybe some other features like how it is spinning. The Pauli exclusion principle basically says you can't have two identical fermions doing exactly the same thing in exactly the same place.

Transferring vibrations

The idea that matter particles are discrete vibrations in fermionic fields helps explain features of the real world that would otherwise be puzzling, such as how particles can be created and destroyed. Back in the heady early days of quantum mechanics, people were struggling to understand the phenomenon of radioactivity. They could see how photons could be created from other particles, because those were just vibrations in the electromagnetic field. But what about radioactive processes, like the decay of the neutron? Inside a nucleus, huddled in close comradeship with a few protons, a neutron can last forever. When it is isolated by itself, however, a neutron will decay within a matter of minutes, transforming into a proton by emitting an electron and an antineutrino. The question is, where did that electron and antineutrino come from? People speculated that they had actually been hidden inside the neutron all along, but that didn't seem quite right.

A beautiful answer was worked out in 1934 by Enrico Fermi, in the first real application of field theory to fermions—which was only appropriate, since those particles had been named after Fermi in the first place. Fermi suggested that you could think of each of these particles as vibrations in different quantum fields, and that each field exerted a tiny influence on the others, much like playing a piano in one room will cause the strings of a piano in a room next door to gently hum in sympathy. It's not that new particles are magically created out of nothing; it's that the vibrations in the neutron field are gradually transferred to the proton, electron, and antineutrino fields. Because it's quantum mechanics, we can't perceive the gradual transfer; we observe the neutron, and we either see it as a neutron, or we see that it's decayed, with some probability that can be mathematically calculated.

Quantum field theory also helps understand how one particle can convert into others that it doesn't even interact with directly. A classic example, and one that will be very important for us very soon, is a Higgs boson decaying into two photons. That sounds surprising, because we know that photons don't couple directly to the Higgs. Photons couple to charged particles, and the Higgs couples to massive particles—and the Higgs is not charged, and photons are not massive.

The trick is the concept of virtual particles, which really should be thought of as virtual fields. A Higgs boson comes along, a vibrating wave in the Higgs field. That vibration can set up vibrations in the massive particles that the Higgs couples to. But maybe those vibrations don't quite rise to the level of appearing as new particles; instead, they set up vibrations in yet another kind of field, in this case the electromagnetic field. That's how a Higgs can turn into photons: First it turns into virtual charged, massive particles, and then they quickly convert into photons.

It's as if you had two pianos that were completely out of tune with respect to each other, and ordinarily wouldn't resonate at all; but there's a third instrument in the room, like a violin, that has enough flexibility to resonate with both of them.

Conservation laws

Because all particles arise from fields, even matter particles can appear and disappear in nature. But it's not as if chaos has completely broken loose. Count up the electric charge before and after the neutron decays. Beforehand it's zero, because you just have a chargeless neutron. Afterward it's also zero; the proton has a positive charge, but the electron has a precisely balancing negative charge, and the antineutrino has no charge at all. It also seems that the number of quarks is the same before and after, since a single neutron produced a single proton. Finally, the number of leptons is exactly one before and after, if we introduce a trick of counting antimatter leptons as "minus one lepton" (and antiquarks as "minus one quark," had there been any of those). Then the neutron is three quarks and zero leptons, while its decay products also add up to three quarks (the proton) and zero leptons (one for the electron and minus one for the antineutrino). That's the reason we know an antineutrino rather than a neutrino is produced when neutrons decay.

These patterns are *conservation laws*—unbreakable rules that govern what particle interactions are allowed in nature. Along with the famous law of conservation of energy, we also have conservation of electric charge, of the number of quarks, and of the number of leptons. Some conservation laws are more inviolable than others; physicists suspect that quark and lepton numbers can sometimes change (very rarely, or under extreme conditions), but most believe that energy and electric charge are absolutely fixed.

With these rules in mind, we can understand which particles decay, and which ones last forever. The rule of thumb is that heavy particles like to decay into lighter ones, as long as the decay doesn't violate any conservation laws. Electric charge is conserved, and electrons are the lightest charged particles, so they are completely stable. Quark number is conserved, and the proton is the lightest particle with nonzero quark number, so it is also stable (as far as we know). Neutrons are not stable, but they can form stable nuclei in the company of protons.

The Higgs boson, a very heavy particle with zero charge that is neither a quark nor a lepton, decays extremely quickly, so fast that we will never observe it directly in a particle detector. That's one of the reasons it's been so hard to find, and why our apparent success has been so sweet.

EIGHT

THROUGH A BROKEN MIRROR

In which we scrutinize the Higgs boson and the field from which it springs, showing how it breaks symmetries and gives the universe character.

In an otherwise empty seminar room at the California Institute of Technology, I was seated on one side of a table and local TV reporter Hal Eisner was seated across from me. In between us was a giant bowl of popcorn. Eisner seized a kernel of popcorn and waved it in front of my nose, asking me—begging, really—to use it to explain the Higgs boson. "If there were no Higgs boson, would this popcorn explode? It would explode, wouldn't it?"

It was September 10, 2008, the day the first protons circulated around the LHC. For a previous generation of accelerators, startup was an understated affair, watched closely by a small band of interested physicists and ignored by the rest of the world. But the LHC is special, and the attention of people worldwide was focused on a handful of protons working up the strength to travel all the way around a seventeen-mile ring for the first time.

Hence, the reporters had come to Caltech, and other universities in other cities, to report on the excitement. It was early morning Geneva time, but California is nine hours behind, so it was late the previous night for us. Computer monitors were set up for everyone to follow along, although the strain on CERN's servers soon broke the Internet feed. Pizza was ordered and passed around, helping the assembled scientists settle into a comfort zone. (A substantial fraction of the atoms in the body of a typical physicist were once in the form of pizza.)

Still, the local news folks were very reasonably asking, what's the big deal? We know this is important, but why, exactly? The search for the Higgs boson was always one of the first answers offered. Okay, so why is the Higgs so important? Something about mass, and breaking symmetries. Let's get down to brass tacks: *Would the popcorn explode?*

The right answer is "yes, if the Higgs boson (or more carefully, the Higgs field in which the boson is a propagating wave) were to suddenly disappear, ordinary matter would no longer hold together, and objects like kernels of popcorn would immediately explode." But it is misleading to think of the Higgs as some kind of force that binds atoms together. The Higgs is a field that permeates space, giving heft to particles like electrons, allowing them to form atoms, which bind into molecules. Without the Higgs, there wouldn't be atoms, there would just be a bunch of particles zooming separately through the universe.

It's a common problem when translating deep concepts from modern physics into the language of everyday life. You want to say things that are completely correct (of course), but you also want to give people the right *impression*, which isn't the same thing—it does no good to say correct things if nobody has a clue what you're talking about, and they might even start thinking something wrong on the basis of your explanation.

Fortunately for us, it's not *that* hard to really understand what's going on. The Higgs field is like the air, or the water for fish in the sea; we don't usually notice it, but it's all around us, and without it life would be impossible. And it is literally "all around us"; unlike all the other fields of nature, the Higgs is nonzero even in empty space. As we move through the world, we are embedded in a background Higgs field, and it's the influence of that field on our particles that gives them their unique properties.

The Higgs boson is not any old particle. When the Tevatron at

Fermilab discovered the top quark in 1995, it was an amazing triumph of effort and ingenuity. But we were already familiar with quarks and weren't really expecting to discover something completely surprising. The Higgs is more than that; we haven't found any other particles like it. Its field fills space, breaks symmetries, gives mass and individuality to the other particles of the Standard Model. If the top and bottom quarks didn't exist, our lives would go on essentially unchanged. If the Higgs boson didn't exist, the universe would be an utterly different place.

A prizewinning analogy

In 1993, the LHC was still an idea on a drawing board, and it was far from certain that it would make the journey to reality. A group of physicists from CERN were pitching the massive project to William Waldegrave, the science minister of the United Kingdom at the time. Waldegrave was interested in the idea, but he couldn't quite grasp the central selling point: the idea of the Higgs boson. "He didn't understand a word of what was said," recalled physicist David Miller of University College London.

But Waldegrave didn't simply give up; he challenged the scientists to provide him with an understandable explanation of the role of the Higgs boson, one that would fit on a single piece of paper. He offered a bottle of vintage champagne to whoever came up with the best explanation. Miller and four colleagues managed to cook up an engaging metaphor that was deemed suitable by the science minister. All five got bottles of champagne, and of course the United Kingdom supported the LHC.

Here's an updated version of Miller's analogy. Imagine that Angelina Jolie and I both walk across an empty room. (The original explanation used Margaret Thatcher rather than a movie star, for obvious political reasons, but all that matters is that we consider someone famous.) For purposes of the thought experiment, let's assume that the speeds at which we naturally walk are the same. In that case, we will cross the room in the same amount of time. There is a symmetry: It doesn't matter whether it's Angelina or me who is walking across the room; the elapsed time will be equal.

Now imagine that there is a party going on in the room, and it's filled with revelers chatting away. I walk across the room, maybe a bit more slowly than I did when it was empty: I have to briefly pause and adjust my path to weave through all the partygoers, but for the most part I pass through unnoticed. When Angelina walks across the same room, it's a completely different story. As she passes by, all sorts of people stop her to get autographs or take pictures or just make small talk. Effectively, her "mass" is larger: It takes more effort for her to get moving and cross the room than it would for me. (I am not saying that Angelina Jolie is fat; it's just a metaphor.) The symmetry that we used to have is broken by the presence of other people in the room.

A physicist would say that Angelina Jolie "interacts more strongly" with the party guests than I do. That strength of interaction is a reflection of her greater celebrity; nobody thinks to stop me and get an autograph, but a famous actress undergoes frequent interactions with the background crowd.

Now replace me with an up quark, Angelina with a top quark, and the partygoers with the Higgs field. If there is no Higgs field filling space, there is a perfect symmetry between an up quark and a top, and they behave in the same way, just as Angelina and I walk across an empty room at the same speed. But a top quark interacts more strongly with the Higgs than an up quark does. If the Higgs field is "turned on," the top gets a greater mass, and it takes more effort to get it moving, just like it takes Angelina more effort to push through a crowd of partygoers than it takes me.

As with any analogy, this one is not perfect. Like a crowd of partygoers, the Higgs field fills space, affecting anything that moves through it. But unlike a crowd of people, or anything else we are familiar with, I can't measure my velocity with respect to this background field; it looks exactly the same no matter how I am moving. It takes more effort to get a particle moving in the presence of the Higgs field, but once it gets moving it stays moving, just as Galileo or Newton or Einstein would have expected. The Higgs field doesn't drag you down to its velocity, because it doesn't have a velocity. There's really no analogy for that in everyday life, but it's how the world appears to work.

Before Einstein and relativity came along, many physicists thought that the electromagnetic waves were vibrations in a medium called the "aether." They even tried to detect the aether by looking for changes in the speed of light depending on the motion of the earth; if the light was traveling in the same direction as the aether it should move faster, and against the aether it should move slower. But they found no difference. It was the genius of Einstein to realize that the whole idea of aether was unnecessary, and the speed of light is absolutely constant through empty space. You don't need an aether field to support the electromagnetic field; the electromagnetic field can just exist.

It's tempting to think of the Higgs field as similar to the aether—an invisible field through which waves move, but the waves are of Higgs bosons rather than electromagnetic radiation. It's not completely inaccurate, since the Higgs field does fill space, and Higgs bosons are vibrations within it. But for the most part this is a temptation to be resisted. The whole point of the aether was that it did matter how quickly you moved within it—it defined a state of rest for empty space. Whereas with the Higgs field, it makes no difference at all. Relativity still works.

Pushed away from zero

As we learned in the last chapter, the universe is made of fields. But most of these fields are turned off—set to zero—in empty space. A particle is a little vibration in a field, a bundle of energy that is created when the field is nudged away from its natural value. The Higgs is different; even in empty space, it's not zero. The field takes on a certain steady value absolutely everywhere, and the Higgs boson particle is a vibration around that value, rather than a vibration around zero. What makes the Higgs so special?

It all has to do with energy. Think of a ball at the top of a hill. It has what physicists call "potential energy"—it's not doing anything, just sitting there peacefully, but it has the potential to release energy if we let it roll down the hill. When that happens, it picks up speed, gradually turning its potential energy into energy of motion. But it also bumps into other rocks, feels air resistance, and makes noises as it moves, all of which dissipate energy along the way. By the time it reaches the bottom of the hill, its original energy has been turned into sound and heat, and the ball can come to rest.

Fields are like that. If we push them away from their preferred resting state, we give them potential energy. Let them go, and they start vibrating, and they can ultimately dissipate their energy by transferring it to other fields. Eventually they settle back to sitting at rest. What makes the Higgs field special is that its resting place isn't at zero at all—its lowest energy state has the field stuck at a value of 246 GeV. That's a number that we determine from experiment, since it determines the strength of the weak interactions.

This number 246 GeV isn't the mass of the Higgs boson (which is about 125 GeV, and was unknown until the LHC found it), it's the value of the field in empty space. Particle physicists like to measure everything in the same units of GeV, which can get confusing. The mass of the Higgs boson tells us how much force we need to get it moving when we push it, just like the mass of any other object; said another way, it's how much energy we need to put into a vibration of the field before it appears to us as a discrete particle. The value of the field is something completely different, characterizing what the field is doing when it's sitting completely still.

To get a handle on why the Higgs field sits near 246 GeV rather than near zero, think of a pendulum suspended from the ceiling. This behaves like a regular field; its lowest-energy state is when it's pointing straight down, sitting still at the bottom of its arc. We can give it energy

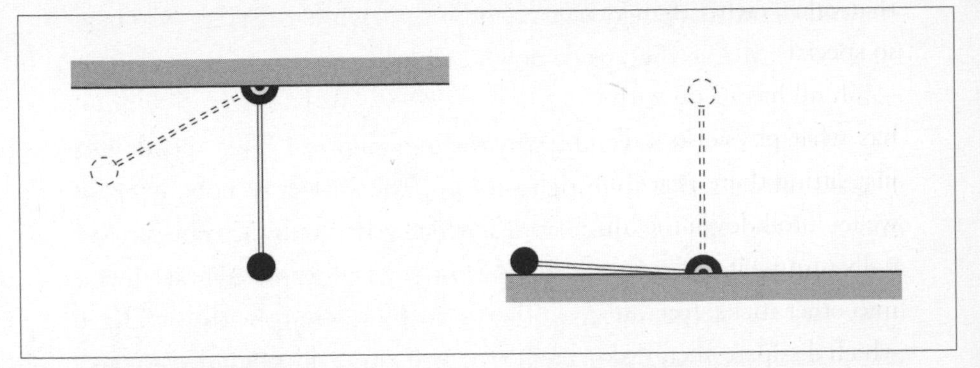

An ordinary field is like a pendulum suspended from the ceiling. It has the least possible energy when it's pointing straight down, at rest. We can pull it up, but that requires energy. The Higgs field is like an upside-down pendulum, stuck to floor rather than suspended from the ceiling. Now it would take energy to lift it to an upright position; the state of lowest energy has the pendulum on the floor, either to the left or to the right.

by pushing it away from that position; if we let it go it will start oscillating back and forth, eventually settling down because it loses energy to air resistance and friction.

Next imagine an upside-down pendulum, one whose pivot is attached to the floor rather than the ceiling. It's the same basic mechanism, but now it behaves completely differently. The inverted pendulum has energy when it's pointed vertically, whereas before that was its lowest-energy configuration. Now there are actually two lowest-energy possibilities: one where it is resting on the floor to the left, and one where it is resting on the floor to the right. Left to its own devices, the pendulum will sit on the floor, pointed left or right.

The Higgs field is similar to the upside-down pendulum, in that it actually costs energy to be at zero. Its lowest-energy state is one in which the field takes some fixed value everywhere, just like the end of the pendulum sits at some distance to the left or right of the pivot. This is why empty space is filled with the Higgs field, through which other particles move and pick up mass: because that's the configuration of lowest energy. The value of the field is like the displacement of the pendulum from vertical; an ordinary field wants to be at zero, while the Higgs wants to be offset, just like the upside-down pendulum wants to point left or right.

Of course we can wonder why the metaphorical Higgs pendulum is upside down rather than right-side up. The answer is, nobody really knows. There are some speculations, which rely on physics way beyond the Standard Model, but at the present state of knowledge it's just a brute fact about the universe. There's nothing wrong with the Higgs taking on a nonzero value in empty space; it either does or it doesn't, and it turns out that it does. And good thing, because otherwise the world would be a lot more boring (and not just for particle physicists).

Giving particles mass

It wouldn't matter that the Higgs field filled empty space—indeed, we wouldn't even notice—if it didn't interact with other particles. And the most obvious effect of that interaction is to "give mass" to the elementary particles of the Standard Model. But this concept is sufficiently subtle that it's worth our time thinking about it a bit. For even more details about how this works, see Appendix One.

First and foremost, we should say what the "mass" of an object is. Probably the best way of thinking about it is "how much resistance you feel when you push on the object," which is another way of saying "how much energy you need to get the thing moving to a certain speed." A car has a lot more mass than a bicycle, which we know because it takes a lot more work to push a car than to push a bike. Another definition would be the "amount of energy an object has while at rest." That's working backward from Einstein's $E = mc^2$. We usually think of this equation as telling us how much energy there is in an object with a certain mass; equivalently, we can think of it as a definition of the mass of an object that isn't moving.

It's important to emphasize that mass is not directly related to gravity at all. We tend to associate the two, because the easiest way to measure the mass of something is to weigh it by putting it on a scale, and we all know that it's gravity that's pulling us down on the scale. Out in empty space where gravity isn't important, things become weightless, but they still have mass. It is harder to get a massive rocket ship moving than a tiny pebble, and it would be harder still to push the moon or a planet. Gravity is something different, which affects all forms of energy, even ones that have no mass. Light, which consists of massless photons, is definitely affected by gravity, as has been vividly demonstrated by the phenomenon of gravitational lensing (bending of light rays) by galaxies and dark matter in the universe.

If you take a look at the Particle Zoo table in Appendix Two, you'll see that some particles have mass and some don't. Among the force-carrying bosons, the gluons, graviton, and photon all have zero mass, but the W and Z bosons do have mass, as does the Higgs itself. Under the fermions, we see that the neutrino masses are listed as "small," while the quarks and charged leptons have specific masses.

This messy situation is ultimately due to the influence of the Higgs field. The rule is simple: If you don't interact directly with the Higgs, you have zero mass; if you do interact directly with the Higgs, you have a nonzero mass, and your mass is directly proportional to how strong that interaction is. Particles like the electron and up-and-down quarks interact with the Higgs boson relatively weakly, so their masses are small; the tau lepton and the top-and-bottom quarks interact with it strongly, so their masses are relatively large. (The neutrinos are a special case; they have tiny masses, but our understanding of where those masses come from is still far from settled. For the most part we'll be ignoring them in this book, and sticking to the parts of the Standard Model that we understand.)

If the Higgs were like other fields, resting at zero in empty space, its interaction strength with other particles would simply measure how likely it would be for the Higgs boson to interact with that particle if they happened to pass by each other. Mostly a Higgs and an electron would pass by in peace, while a Higgs and a top quark would scatter very strongly. (I can pass by strangers on the street without being interrupted, but Angelina Jolie would be hassled at every step.) But because the expectation value is not zero, it's like the other particles are interacting with it *constantly*—and it's those persistent, inevitable interactions with the background that create the mass of the particle. When a particle interacts strongly with the Higgs, it's as if it carries a large crowd of Higgs hangers-on everywhere it goes, contributing to its mass.

The formula for the mass of a particle is pretty easy: It's the value of the Higgs field in empty space, times the particular interaction strength that the particle has with the Higgs. Why do some particles, like the top quark, interact strongly with the Higgs, and others, like the electron, interact relatively weakly? And what explains the specific numbers? Nobody knows. Right now, these are unanswered questions. At the current state of the art, we treat those coupling strengths as constants of nature that we simply have to go out and measure. We're hoping to get some clues by studying the Higgs itself, which is one reason the LHC is so important.

A world without Higgs

Despite all that, it's misleadingly sloppy to say, "The Higgs is responsible for mass," as we physicists sometimes do. Remember that we don't see the quarks directly; they are confined, along with the gluons, inside hadrons such as protons and neutrons. The mass of a proton or neutron is much greater than the masses of its individual quarks, and for good reason; it mostly comes from the energy of the virtual particles that are binding the quarks together. If there were no Higgs, quarks would still bind together to form hadrons, whose masses would be practically unchanged. This means that most of the mass of, say, a desk, or a person, doesn't come from the Higgs boson at all. The large majority of the mass of ordinary objects comes from their protons and neutrons, and that comes from the strong interactions, not from the Higgs field.

Which isn't to say that the Higgs is irrelevant to everyday physics. Imagine we got our hands on a secret control panel that governed all the laws of physics, and by slowly turning the dial labeled HIGGS we could decrease the value of the Higgs field in empty space from 246 GeV to any smaller number. (Note: There is no such secret panel.) As the background Higgs field all around us diminished in value, so would the masses of the quarks, the charged leptons, and the W and Z bosons. The changes in the masses of the quarks and W and Z bosons would lead to tiny changes in the properties of protons and neutrons, but nothing immediately dramatic. The changes to the muon and tau are basically irrelevant to everyday life. But any change in the mass of the electron would be hugely significant.

In our usual mental cartoon picture of an atom, electrons orbit around the nucleus just like planets orbit the sun or the moon orbits the earth. This is a case where the cartoon breaks down, and we have to take quantum mechanics seriously. Unlike a planet orbiting the sun, a typical electron isn't orbiting at some random distance; it's actually going to be as close to the nucleus as it can possibly get. (If it's farther away, it will tend to lose energy by giving off a photon and therefore move closer.) And how close it can get depends on its mass. Heavy particles can squeeze into small regions of space, while lighter particles are always more spread out. The size of atoms, in other words, is determined by a fundamental parameter of nature, the mass of the electron. If that mass were less, atoms would be a lot larger.

That's a big deal. If we made atoms bigger, it's not as if the size of ordinary objects would grow along with them. What makes ordinary stuff hang together is chemistry—the ways in which atoms stick together in interesting combinations. And the reason they stick together is because they share electrons, at least under the right circumstances. And those circumstances would completely change if atoms had different sizes. If the mass of the electron changed just a little bit, we would still have things like "molecules" and "chemistry," but the specific rules that we know in the real world would change in important ways. Simple molecules like water (H_2O) or methane (CH_4) would be basically the same, but complicated molecules like DNA or proteins or living cells would be messed up beyond repair. To bring it home: Change the mass of the electron just a little bit, and all life would instantly end.

Change the mass of the electron by a lot, and the effects would be correspondingly more dramatic. As the Higgs field got closer and closer to zero, electrons would get lighter and lighter, and atoms would get correspondingly bigger. Eventually they would reach macroscopic size, and then astronomical size. Once every atom is as big as the solar system, or the Milky Way galaxy, there's no sense in talking about "molecules" anymore. The universe would just be a collection of individual super-enormous atoms, bumping into one another in the cosmos. If the electron mass were turned all the way down to zero, there wouldn't be atoms at all—the electrons wouldn't be able to stick to the nuclei. And if that happened suddenly, Hal Eisner's leading question would be answered—the popcorn kernel would explode.

There is something more subtle going on, as well. Think of the three charged leptons: the electron, the muon, and the tau. The only differences between these particles are their masses. If we turn off the Higgs field, those masses go to zero, and the particles become identical. (Technical aside: The strong interactions can also give fields expectation values, mimicking the effects of the Higgs but at a much lower value; we're ignoring those in this discussion.) The same holds true for the three quarks with charge +2/3 (up, charm, and top) and for the three quarks with charge −1/3 (down, strange, and bottom). Each group of particles would be identical if it weren't for the Higgs background. This points to perhaps the most basic role of the Higgs field: It takes a symmetric situation and breaks it.

Defining symmetry

When we think of the word "symmetry," what comes to mind is a pleasing regularity. Studies have shown that symmetric faces, ones that look the same on the left and the right, are generally found to be more attractive. But physicists (and the mathematicians from whom they learn things like this) want to go deeper, studying what makes something "symmetric" in the most general sense, and how those symmetries appear in nature.

The simple notion of "matching left and right sides" reflects a broader idea: We say that an object possesses a symmetry whenever we can do something to it and be left with exactly what we started with. For a symmetric face, we can imagine reflecting it around a line down the middle and getting back the same face. But simpler objects can have much more symmetry than that.

Think of a geometric figure like a square. We can take its mirror image, reflecting both sides of the square around a vertical axis drawn precisely down the middle, and get back exactly the square we started with—that's a symmetry. We could also do the same thing around a horizontal axis, which indicates an additional symmetry. (That wouldn't

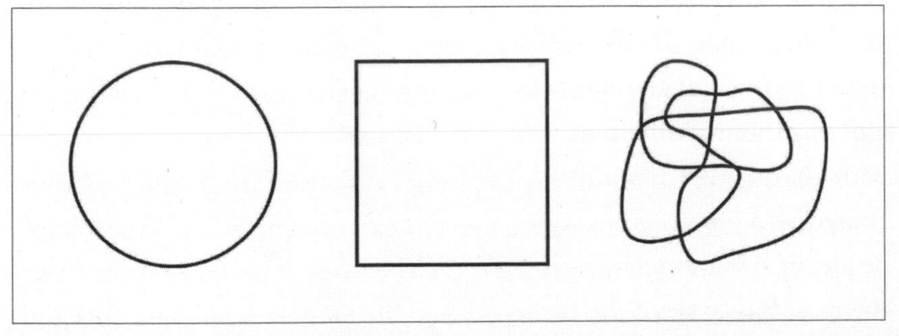

A circle, a square, and a scribble. The circle has a great deal of symmetry, including rotations of any angle and reflections around any axis. The symmetries of the square are fewer: rotations by ninety degrees, reflections around vertical or horizontal axes, or combinations thereof. The scribble has no symmetry at all.

have worked with a face; even the most beautiful person looks different when seen upside down.) For that matter, we could reflect about either diagonal axis—but not a random axis, which would move the corners of the square around. We can also rotate the square clockwise around its center by ninety degrees, or any multiple thereof.

A circle, like a square, looks very symmetric, and in fact it's much more so. We cannot only reflect it around any axis through the center, we can rotate it by any angle whatsoever, and it will always come back to an identical-looking circle. That's much more freedom than we had with the square. A random scribble, by contrast, doesn't have any symmetry at all. Any way in which we alter it will leave it looking different.

A symmetry is a way of saying "we can alter things in some particular way and nothing important changes." It doesn't matter if we rotate the square by ninety degrees, or reflect it about a central axis: It ends up looking the same.

From this perspective, the idea of symmetry might not seem that powerful. So it doesn't matter if we rotate the circle; who cares? The reason we care is because sufficiently powerful symmetries place very strong constraints on what can possibly happen. Suppose someone tells you, "I have drawn a figure on this piece of paper, with so much symmetry that you can rotate the paper by any angle and the figure will look the same." Then you know that the figure has to be a circle (or a single point, which is sort of a circle of zero size). That's the only figure that has so much symmetry. Likewise, when it comes to physics, we can often figure out how experiments should behave just by understanding that there is an underlying symmetry at work.

A classic case of symmetry in physics is the simple observation that it doesn't matter where we do a certain experiment; if the experiment reflects basic underlying principles, we will get the same result. For example, there is a famous experiment in which a scientist (usually young, and often filmed for later YouTube consumption) introduces Mentos candies into a bottle of Diet Coke. The porous structure of the mints helps to catalyze the release of carbon dioxide from the soda, resulting in an impressive geyser of foam. The experiment doesn't work as well with other kinds of candies, or other kinds of soda; but it works exactly the same when carried out in Los Angeles, Buenos Aires, or Hong Kong. There is no symmetry of nature under the interchange of different kinds of food or drink, but there is a symmetry of changing position. Physicists call this "translation invariance," because they can't resist the opportunity to give an intimidating name to a simple concept.

When it comes to particles or fields, symmetries tell us that we can exchange different kinds of particles, or even "rotate them into each other." (Scare quotes are useful here because we're transforming fields into each other, not rotating directions in the honest three-dimensional space in which we live.) The most obvious example is the three kinds of colored quarks, conventionally labeled "red," "green," and "blue." Which label is which is completely irrelevant—if you have three quarks in front of you, it doesn't matter which one you call the "red quark" and which one you call the "blue quark" and which one you call the "green quark." You can change those labels and all the important physics remains unaltered—that's the power of the symmetry. If you had one quark and one electron, you wouldn't want to switch their labels. A quark is very different from an electron; it has a different mass, a different charge, and it feels the strong interaction. There's no symmetry at work there.

If it wasn't for the Higgs field giving masses to the elementary particles, there would be a symmetry that related the electron, muon, and tau, since those particles would be identical in every way, just as Angelina and I moved at equal speeds through the empty room. We could switch a muon in for an electron in some interaction, and the details would be the same. We could even (according to the rules of quantum mechanics) make a particle that was half-electron and half-muon, and it would also be identical, or for that matter any combination of the three particles—much like we can rotate a circle by any angle. Similar symmetries would apply to the up/charm/top quarks, as well as to the down/

strange/bottom quarks. These are known as "flavor" symmetries, and even though the Higgs prevents them from being perfectly respected in nature, they remain very helpful to particle physicists analyzing different basic processes.

But there's another symmetry, deeper and more subtle than the flavor symmetries, that seems completely hidden at first but turns out to be of absolutely crucial importance. That's the symmetry underlying the weak interactions.

Connections and forces

The real importance of symmetries—the reason why physicists can't stop talking and thinking about them—is that sufficiently powerful symmetries give rise to forces of nature. That's one of the most astonishing insights of twentieth-century physics, but it's not an easy one to grasp. It's worth going down the rabbit hole just a bit to understand how symmetries and forces are connected.

Just as there is a symmetry of the everyday world that says "it doesn't matter where you do your experiment," there is another one that says "it doesn't matter in which direction your experiment is pointing." Put the Mentos in the Diet Coke and watch the foam fly; then rotate the whole apparatus from facing north to facing east, do it again, and (within experimental uncertainties) you should get the same result. This is called "rotational invariance," for obvious reasons.

In fact it goes further than that. Let's say I'm doing my experiment in the parking lot outside my office, and a friend is doing another experiment a few feet away, completely unconnected to mine. We could both rotate our equipment by some angle and expect to get the same results. But even better, I can rotate my equipment and she could keep hers just as it was, or we could both rotate by some arbitrary angle. In other words, the symmetry is not just a single rotation of the world (it doesn't matter whether we're all facing north, or some other direction), but separate rotations at every single point (it doesn't matter what direction any of us is individually pointing in).

That's an enormously larger amount of symmetry. In the trade this kind of megasymmetry is called a "gauge invariance." The name was given by German mathematician Hermann Weyl, who likened the choice of how to measure things at different points to the choice of gauge (distance between rails) in railroad tracks. They are also called "local" symmetries, since we can do the symmetry transformation separately at every location. A "global" symmetry, by contrast, would be based on a transformation that must be carried out uniformly everywhere at the same time. (Local doesn't mean "only at one point"; it means "separately at every point." Local symmetries are bigger and more powerful than global symmetries.)

Because we can set up our equipment in different directions at every point, it becomes crucial that we can somehow compare the actual setup we choose at different points. Think of surveyors, laying out the plans for a new house. They can start with one corner, which fixes the direction in which the house will be oriented. But, presuming the house has the shape of a rectangle, they're going to want the orientation of the other corners to line up with the first one; you can't have the bricks at the four corners of your house just pointing in random directions. In the real world, this usually isn't too hard; we simply need to draw some straight lines, either by pulling string between the points or through the use of surveying equipment.

Imagine, however, that the ground on which we're building our house isn't completely level. The terrain is bumpy, and for aesthetic reasons our client wants us to build on top of them rather than just bringing in the bulldozers and leveling the place. In that case, our problem becomes a little trickier; we need to take the variations of the ground into account when we figure out how to line up the corners of our building.

Here's the subtle point: The way we connect our notions of "the same direction" at different points in space requires that there is a field filling the space between those points—a field that literally tells us how to connect them together, and in the technical literature is called a "connection." In our architectural example, the relevant field comes from the height of the ground itself. That's a field—it's not a fundamental field that vibrates to give particles, but it's a number at every point along the ground, which is all a field really is. (A topographical map would be a picture of the "height field.") The information in that field lets us relate what happens at different points in space.

Whenever we have a symmetry that allows us to do independent transformations at different points (a gauge symmetry), it automatically comes with a connection field that lets us compare what is going on at those locations. Sometimes the field is completely innocuous and doesn't even get noticed, like the height of the ground on a surface that is perfectly flat. But when the connection field twists and turns from place to place, it has enormous consequences.

For example, when the height of the ground changes from place to place, you can go skiing on it (or skateboarding, depending on the conditions). If the ground is flat, you would just sit there unmoving; when the ground is sloped, there is a force that pulls you down the hill. That's the magic formula that makes the world go, according to modern physics: Symmetries lead to connection fields, and bends and twists in the connection fields lead to forces of nature.

The four forces of nature—gravity, electromagnetism, and the strong and weak nuclear forces—are all based on symmetries. (The Higgs boson

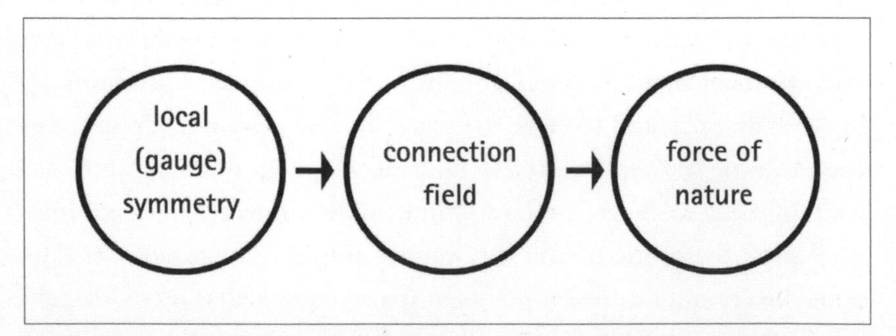

Where the forces of nature come from: Local symmetries imply the existence of connection fields, which give rise to forces.

also carries a force, but it's not what gives particles masses—that's the Higgs field in the background. And it's not based on any symmetry.) The boson fields that carry those forces—gravitons, photons, gluons, and the W and Z bosons—are all connection fields that relate those symmetry transformations at different points in space. They are often called "gauge bosons" to drive home the point.

The connection fields define invisible ski slopes at every point in space, leading to forces that push particles in different directions, depending on how they interact. There's a gravitational ski slope that affects every particle in the same way, an electromagnetic ski slope that pushes positively charged particles one way and negatively charged particles in the opposite direction, a strong-interaction ski slope that is only felt by quarks and gluons, and a weak-interaction ski slope that is felt by all the fermions of the Standard Model, as well as by the Higgs boson itself.

For gravitons, the symmetries responsible for the force are the ones we've already talked about—translations (changes of position) and rotations (changes in orientation)—but in four-dimensional spacetime, not just three-dimensional space. For the strong interactions, the symmetry relates the colors red, green, or blue of the different quarks. It doesn't matter whether we describe a certain quark as red, green, blue, or any combination thereof, so that's a symmetry.

You might have noticed that particles with electric charge always come in matched pairs: one with a positive charge, and one with a negative charge. That's because, to get a charged particle, you need two fields that can rotate into each other under the gauge symmetry of electromagnetism. A single field by itself can't be electrically charged, since there's nothing for the symmetry to act on.

This leaves us with the W and Z bosons of the weak interactions. They are also connection fields, born out of a certain underlying symmetry of nature. But that symmetry is masked by the Higgs field, so it takes a bit more work to describe.

The problem with symmetries

The symmetry underlying the weak interactions was discovered in a roundabout fashion. Back in the 1950s, before the idea of quarks had even been invented, physicists had noticed that neutrons and protons were pretty similar in some ways. The neutron is a tiny bit heavier, but its mass is close to that of the proton, all things considered. Of course the proton has an electric charge and the neutron doesn't, but the electromagnetic interaction isn't as strong as the strong nuclear force, and as far as the strong force goes, the two particles seem indistinguishable. If we were interested in the strong interactions in particular, we could make a lot of progress by thinking of the neutron and proton as just two different versions of a unified "nucleon" particle. That's at best an approximate symmetry—the charges and masses really are different, so the symmetry isn't perfect—but you can still squeeze a lot of usefulness out of it.

In 1954, Chen Ning Yang and Robert Mills came up with the idea that this symmetry should be "promoted" to a local symmetry—i.e., that we should be allowed to "rotate" neutrons and protons into each other at every point in space. They knew what this implied: the existence of a connection field and a corresponding force of nature. At face value, it might have seemed like a somewhat crazy idea; how do you make a gauge symmetry out of something that is only approximately a symmetry in the first place? But it often happens that crazy ideas are later recognized as brilliant ones as we understand more about how nature works.

There was a bigger problem. At the time, there were two successful theories based on local symmetries: gravity and electromagnetism. You'll notice that they are both long-range forces, and that the bosons that mediate the forces have zero mass. Neither of these facts is a coincidence. It turns out that the requirement of local symmetry demands that the associated boson be exactly massless; and when you have a massless boson, the force it carries can extend over very long ranges. The force from a massive boson peters out quickly due to the energy required to make the massive particles, but the force from a massless one can reach out indefinitely far.

The thing about massless particles is they're easy to make. Especially if we are talking about a field that interacts readily with neutrons and protons, and trying to understand what happens inside an atomic nucleus, where the forces are clearly very strong. From the 1954 point of view, it seemed obvious that there weren't any new massless particles playing an important role inside the nucleus. But Yang and Mills persevered.

It wasn't easy. In February of that year, Yang gave a seminar at the Institute for Advanced Study at Princeton on his new work. In the audience, among other luminaries, was the famously acerbic physicist Wolfgang Pauli. Pauli knew perfectly well that the Yang-Mills theory predicted a massless boson, in part because Pauli himself had investigated a very similar model but never published. He wasn't the only one; other physicists, including Werner Heisenberg, contemplated similar ideas before Yang and Mills put it together explicitly.

As an audience member in a scientific seminar, it may occasionally happen that you disagree with something the speaker is saying. The usual protocol is to ask a question, perhaps make a statement to register your disagreement, and then let the speaker continue. That wasn't Pauli's style. He interrupted Yang repeatedly, demanding to know, "What is the *mass* of these bosons?"

Yang, who had been born in China in 1922 and had moved to the United States to study with Enrico Fermi, would share the 1957 Nobel Prize with T. D. Lee for their work on the violation of parity (left-right symmetry). But just a few years earlier he was still relatively young and not yet established. In the face of Pauli's onslaught, Yang found himself at a loss, and eventually he simply sat down quietly in the middle of his own seminar. Robert Oppenheimer, who was chairing the proceedings, coaxed him into resuming his talk, and Pauli stewed in silence. The next day, Pauli sent a simple note to Yang: "I regret that you made it almost impossible for me to talk to you after the seminar. All good wishes. Sincerely, W. Pauli."

Pauli wasn't wrong to worry about the prediction of unseen massless particles, but Yang wasn't wrong to pursue his idea despite this apparent flaw. In their paper, Yang and Mills admitted the problem but expressed a vague hope that quantum-mechanical effects from virtual particles would give their bosons mass.

They were almost right. Today we know that both the strong interactions and the weak interactions are based on what we call Yang-Mills theories. And the two forces use very different, but equally clever and surprising, ways of hiding their massless particles. In the strong interactions, the gluons are massless, but they're confined inside hadrons, so we simply never see them. In the weak interactions, the W and Z bosons would be massless if it wasn't for the interference of the Higgs field pervading space. The Higgs breaks the symmetry on which they are based, and once that symmetry is broken there's no reason for the bosons to remain massless. Figuring all that out required quite a journey.

Breaking symmetries

To understand how a symmetry can be "broken," we descend from the land of abstraction back to the everyday world. We've mentioned a couple of simple examples of symmetries around us: It doesn't matter where you are, and it doesn't matter in what direction you are pointing. The laws of physics have another symmetry, but one that's harder to notice: It doesn't matter at what speed you are traveling, an idea first codified by none other than Galileo himself.

Imagine you are on a train, zipping through the countryside. Let's make it a supermodern train, using magnetic levitation to float above the tracks rather than old-fashioned wheels. If the train is sufficiently quiet and free of bumps along the ride, there is no way we can tell what speed we're moving at without looking out the window. Just by minding our business, doing physics experiments inside the train, the speed at which we're moving doesn't matter. We could be completely still or cruising along at 100 miles an hour; the effect of dropping Mentos in the Diet Coke will be exactly the same.

This remarkable fact is hidden from us in our everyday experience, for a simple reason: We can look outside, or just stick our hand out the window. It instantly becomes clear how fast we're moving, because we can measure (or at least estimate) our speed relative to the ground or the air.

This is an example of symmetry breaking. The laws of physics don't care how fast you are going, but the ground and the air definitely do. They pick out a preferred velocity, namely "at rest with respect to the ground." The deep-down rules of the game have a symmetry, but our environment doesn't respect it; we say that the symmetry is broken by the environment. That's exactly what the Higgs field does to the weak interactions. The underlying laws of physics obey a certain symmetry, but the Higgs field breaks it.

The symmetry breaking we've been talking about thus far is often called "spontaneous" symmetry breaking. That's a way of saying that the symmetry is still really there, hiding in the underlying equations that govern the world, but some feature of our environment is picking out a preferred direction. Being able to stick your hand out the window of a train and measure your speed with respect to the air doesn't change the fact that the laws of physics are invariant with respect to different velocities. Indeed, when people are careful they will sometimes talk of symmetries as simply being "hidden" rather than "spontaneously broken." More on this notion of spontaneity in Chapter Eleven.

Symmetries of the weak interactions

It turns out that Yang and Mills were basically on the right track with the idea of a symmetry between neutrons and protons. These days we know about quarks, of course, so the analogous idea would be to propose a symmetry between up quarks and down quarks. The same problems appear to get in the way: the up and down quarks have different masses and different electric charges. If those features can be traced to the existence of the Higgs field, we could be in business. And indeed they can.

Here's where things get messy—so much so, that the details have been relegated to Appendix One. (It's not supposed to be simple; we're talking about a series of discoveries that resulted in multiple Nobel Prizes.) The origin of the messiness resides in the fact that elementary fermions have a property called "spin." Massless particles, which always move at the speed of light, can spin in one of two ways: They can be left-handed or right-handed. Think "spinning clockwise/counterclockwise if the particle is moving toward you." The secret of the weak interactions is that there is a symmetry relating all the left-handed particles, and an associated force, but no matching symmetry for the right-handed particles. The weak interactions violate parity—they discriminate between left and right. You can think of parity as the operation of looking at the world through the reflection in a mirror, where right and left are swapped. Most forces (strong, gravitation, electromagnetism) act the same whether you look at them directly or through a mirror; but the weak interactions treat right and left differently.

The symmetry of the weak interactions relates pairs of left-handed particles, in basically the following way:

up quark ↔ down quark

charm quark ↔ strange quark

top quark ↔ bottom quark

electron ↔ electron neutrino

muon ↔ muon neutrino

tau ↔ tau neutrino

The particles that we've joined up in pairs here seem very different to us at first glance; they have different masses and charges. That's because the Higgs field lurking in the background breaks the symmetry between them. If it weren't for the masquerade put on by the Higgs, the particles in each pair would be completely indistinguishable, just like we think of red/green/blue quarks as three different versions of the same thing.

The Higgs field itself rotates under the symmetry of the weak interactions; that's why, when it gets a nonzero value in empty space, it picks out a direction and breaks the symmetry, just like the air picks out a velocity that we can measure things with respect to when we're traveling in our train. Back in our pendulum example, the lowest-energy state of the regular pendulum was perfectly symmetric, pointing straight down. The upside-down pendulum, like the Higgs field, breaks the symmetry by falling either left or right.

If you were hopelessly lost in a forest in the middle of the night, all directions would seem the same to you. You could rotate how you were standing, and your situation would be just as dire. But if you had a compass, and you knew you wanted to walk north, the direction picked out by that compass would break the symmetry; now there's a right direction to walk, and there are wrong directions. Likewise, with no Higgs field the electron and the electron neutrino (say) would be identical particles. You could rotate them into each other, and the resulting combinations would remain indistinguishable. The Higgs field, like the compass, picks out a direction. There is now one particular combination of fields that interacts most strongly with the Higgs field, which we call the "electron," and one that doesn't, which we call the "electron neutrino." It's only with respect to the Higgs field filling space that such a distinction makes sense.

If it wasn't for the symmetry breaking, there would actually be *four* Higgs bosons, rather than just one; two pairs of particles that transform into each other via the weak interaction symmetry. But when the Higgs field fills space, three of those particles get "eaten" by the three gauge bosons of the weak interactions, which thereby go from being massless force-carriers to being the massive W and Z bosons. Yes, physicists really do talk that way: The weak-interaction bosons gain mass by consuming the extra Higgs bosons. You are what you eat.

Back to the Bang

The analogy between the Higgs field and the upside-down pendulum is actually a pretty good one. Like the Higgs, the underlying laws of physics for the pendulum are perfectly symmetric; they don't favor either left or right. But there are only two stable configurations for the pendulum to be in: pointing left or pointing right. If we tried to balance it carefully so that it was pointing in a symmetric configuration pointing directly upward, any tiny bump would send it falling left or right.

The Higgs field is the same way. It *could* be set to zero in empty space, but that's an unstable configuration. For the pendulum, if it's lying peacefully to the left or right, we would have to exert some energy to lift it so that it pointed directly upward. The same is true for the Higgs field. To move it from its nonzero value at every point in space back to zero would require a superhuman amount of energy—much more than the total energy in the observable universe today.

But the universe used to be a much denser place, with a lot more energy packed into a much smaller volume. At times near the Big Bang, 13.7 billion years ago, matter and radiation were squeezed much closer together, and the temperature was enormously higher. In terms of the pendulum analogy, think of that upside-down pendulum sitting on a table rather than being bolted to the floor. "High temperature" means a lot of random motions of particles; in terms of the analogy, it's like someone takes a hold of the table and starts shaking it. If the shaking is sufficiently energetic, we might imagine that the pendulum is pushed so hard that it flips over from left to right (or vice versa). If the shaking is *really* energetic, the pendulum will vibrate like crazy, flipping quickly back and forth. On average, it will spend as much time on the left as on the right. In other words, at high temperatures, the upside-down pendulum becomes symmetric again.

The same thing happens with the Higgs field. In the very early universe, the temperature is unbelievably high, and the Higgs field is being jostled constantly. As a result, its value at any one point keeps hopping around and averages out to zero. In the early universe, *symmetry is restored*. W and Z bosons are massless, as are the fermions of the Standard Model. The moment at which the Higgs went from being zero on average to some nonzero value is known as the "electroweak phase transition." It's something like liquid water freezing to become ice, but nobody was around to see it happen.

We're talking about very early times in the history of the universe here: about one trillionth of a second after the Big Bang. If you re-created the conditions from the early universe in your living room, the Higgs would evolve from zero to its usual nonzero value so quickly that you'd never notice it had been zero. But physicists can use equations to predict a long sequence of events that happened in that first trillionth of a second. At the moment we don't have any direct experimental data to test those ideas, but we're working on making predictions that will someday confront the observations.

Messy but effective

This story might sound a bit far-fetched, what with nonzero fields in empty space, nature discriminating between left and right, and bosons putting on weight by chowing down on other bosons. It's a picture that was only put together gradually, over the course of many years, and against a tide of skeptical voices chiming in along the way. But . . . it fits the data.

When this theory of the weak interactions was finally put together by Steven Weinberg and Abdus Salam in independent papers from the late 1960s, it was pretty thoroughly ignored. Too much artifice, too many fields doing too many weird things. At the time, people had deduced that something like the W bosons must exist in order to carry the weak force. But Weinberg and Salam predicted a new particle, the neutral Z boson, for which there wasn't any evidence. Then in 1973, an experiment at CERN with the whimsical name of Gargamelle found evidence for the interaction carried by what we now call the Z. (The particle itself wasn't discovered until ten years later, also at CERN.) Since then, experiment after experiment has piled on data that continues to support the basic picture of a weak-interaction symmetry broken by a Higgs field.

As of 2012, we seem to have finally put our fingers on the Higgs itself. But that's not the end of the story, it's only the beginning. There's no question that the Higgs theory fits the data, but in many ways it seems more than a bit contrived. Other than the Higgs, every particle we've ever found is either a fermionic "matter particle" or a boson derived from the connection field associated with a symmetry. The Higgs seems different; what makes it so special? Why just those symmetries, broken in just that way? Is it possible there's a deeper theory that would work even better? Now that we're confronting data rather than just inventing models, there is good reason to hope that we will be inspired to come up with a better theory than brainpower alone has yet given us.

NINE

BRINGING DOWN THE HOUSE

In which we figure out how to find the Higgs boson, and how we know we've found it.

After years of waiting, the discovery of the Higgs boson came faster than anyone had expected.

In one sense, anticipation had been building for more than four decades, since the Higgs mechanism became the accepted model of the weak interactions. But once the LHC started running, excitement grew in earnest in December 2011.

Early that month, CERN had put up a fairly innocuous notice, advertising seminars on December 13 entitled "Update on the search for the Higgs boson by the ATLAS and CMS experiments at CERN." Updates happen all the time, so by itself that wasn't anything to get excited about. But with two experiments, each representing a group of more than three thousand physicists, word quickly spread that these wouldn't be any old seminars. As early as December 1, the British *Telegraph* featured a story by science correspondent Nick Collins, headlined, SEARCH FOR GOD PARTICLE IS NEARLY OVER, AS CERN PREPARES TO ANNOUNCE FINDINGS. The article itself wasn't nearly as breathless as the headline, but the implications were clear. On the physics blog *viXra log*, pseudonymous commenter "Alex" pithily noted, "Today['s] rumour is: Higgs at 125 Gev around 2-3 sigma," leading other commenters to gleefully start speculating about the theoretical implications.

"Alex" could have been anyone, of course, from a mischievous teenager in Mumbai who enjoys tweaking particle physicists to Peter Higgs himself. But multiple blogs and online articles seemed to be pointing in the same direction: This wasn't any old update, this was going to be important news about the Higgs . . . maybe even the long-awaited discovery announcement.

CMS and ATLAS, the two large LHC experimental collaborations, are each miniature republics, in which the citizens elect leaders to represent them. The topmost office is simply called the "spokesperson." To ensure that the collaboration speaks with a unified voice, the preparation and communication of new results is tightly controlled—not only official publications, but even talks by individual collaboration members must be carefully vetted. Talks this important are given by the spokespersons themselves. In December 2011, both spokespersons hailed from Italy: Fabiola Gianotti, a CERN staff member, was the leader of ATLAS, while Guido Tonelli from the University of Pisa headed up CMS.

Gianotti is a major player in experimental particle physics, voted by the *Guardian* as one of the top one hundred women scientists in the world. She came to the field relatively late, as a college student, after concentrating in Latin, Greek, history, and philosophy in high school, and pursuing piano seriously at a conservatory. It was a professor's explanation of the photoelectric effect—Einstein's suggestion that light always comes in discrete quantized packets—that ignited her interest in physics. Now she was leading one of the largest scientific endeavors of all time, on the verge of discovering a major piece of nature's puzzle. Asked to explain the importance of this quest, Gianotti didn't hesitate to use poetic language: "Fundamental knowledge is a little bit like art. It's something very much related to the spirit, the soul, the brain of men and women, as clever beings."

Both speakers had exciting news to report, but they did so in the most cautious manner possible. There were hints. ATLAS, in particular, saw some evidence that looked compatible with a Higgs around 125 GeV. In particle physics, "evidence" for unusual things comes and goes quite frequently, but this wasn't any old unusual thing; it was the kind of signal expected from decaying Higgs bosons, after we had ruled out almost every other place it could be. When you've lost your keys and have searched for them almost everywhere, you shouldn't be surprised when they turn up in the last place you look. To make the case stronger, CMS also saw a wisp of a signal at just about the same mass. Again, nothing to write home about on its own, but in the context of ATLAS's result it was more than enough to get the room buzzing.

Gianotti did her best to bring the enthusiasm under control: "It's too early to tell if this excess is due to a fluctuation of the background, or if it is due to something more interesting." Later she expressed the same sentiment in more colloquial fashion by quoting an Italian saying, "Don't sell the skin until you have caught the bear."

This particular skin was sold and space in the living room was set aside for a nice new rug long before the bear was actually caught. Statistically, the December results might not have been anything to write home about, but they fit perfectly into what physicists expected to see if there was a Higgs at 125 GeV. It seemed just a matter of time before more LHC data would settle the case. It ended up taking less time than we had any right to expect.

What goes in

Let's take a step back and think about what it takes to discover the Higgs boson, or even find tantalizing evidence for its existence. To dramatically oversimplify things, we can boil it down to a three-step process:

1. Make Higgs bosons.
2. Detect the particles that they decay into.
3. Convince yourself that the particles really came from the Higgs, and not something else.

We can examine each step in turn.

We know the basic idea of making Higgs bosons: Accelerate protons to high energy in the LHC, smash them together inside one of the detectors, and hope that a Higgs is produced. There are more details, of course. We can hope to produce the Higgs when we reach very high energies, because $E = mc^2$ tells us that we have a chance of creating high-mass particles. But thinking that there's a chance is different from knowing that it will happen. What are the precise processes by which we can expect to make a Higgs boson?

Your first thought is, "Well, protons smash together, the Higgs comes out." But a little more thought reminds you that protons are made of quarks and gluons, not to mention virtual antiquarks. So it must be that some combination of quarks and gluons smash together to form a Higgs. Then you remember that in Chapter Seven we talked about conservation laws—quantities like electric charge, quark number, or lepton number remain unchanged in any known particle interaction. So we simply can't have, for example, two up quarks smash together and form a Higgs. The Higgs has zero electric charge, while each up quark has charge +2/3, so the numbers don't add up. Adding insult to injury, two up quarks have a total quark number of 2, while the Higgs has a quark number of zero, so that doesn't add up either. If you had a quark and an antiquark come together, you'd have a chance.

What about the gluons? The short answer is, "Yes, two gluons can combine to make a Higgs," but there's a long answer that is a bit more complicated. Remember that the whole point of the Higgs field (or one of the points, anyway) is to give mass to other particles. The more the Higgs interacts with something, the more mass it ends up having. The converse is also true: The Higgs interacts very readily with heavy particles, only reluctantly with light particles, and it doesn't interact directly at all with massless particles like photons and gluons. But through the magic of quantum field theory, it can interact indirectly. Gluons don't interact directly with the Higgs, but they do interact with quarks, and quarks interact with the Higgs; so two gluons can collide to produce a Higgs by going through quarks as an intermediate step.

Particle physicists have developed a very detailed and rigorously tested formalism for understanding how particles interact with one another. Richard Feynman, the colorful Nobel Prize–winning physicist, invented an extraordinarily helpful method for keeping track of these comings and goings: Feynman diagrams. These pictures are little cartoons of particles interacting and evolving over time into other particles. Force-carrying bosons are drawn as wavy lines, fermions are solid lines, and the Higgs is a dashed line. By starting with a fixed set of fundamental interactions, and mixing and matching the corresponding diagrams, we can figure out all the different ways particles can be produced or converted into other particles.

For example, two gluons can come along, represented by wavy lines. These vibrations in the gluon field set up vibrations in the quark fields, which can be thought of as a quark-antiquark pair. Because it's one quark and one antiquark in each case, the total charge and quark number is zero, matching that of the initial gluon. These quarks are virtual particles, playing a crucial intermediary role, but doomed to disappear before they ever show up in a particle detector. One matched quark-antiquark pair meets and they cancel each other out; the other meets and gives rise to a Higgs boson. Every kind of quark contributes to this process, but top quarks contribute the most, since (as the heaviest flavor of quark) they couple to the Higgs most strongly. All of this could be precisely described using a couple of lines of intimidating mathematical machinery; alternatively, it is elegantly captured in a single friendly diagram.

Feynman diagrams provide a fun, evocative way of keeping track of what kinds of things can happen when particles come together to interact. Physicists, however, use them for the very down-to-earth task of calculating the quantum probability of the depicted interaction taking place. Every diagram corresponds to a number, which can be

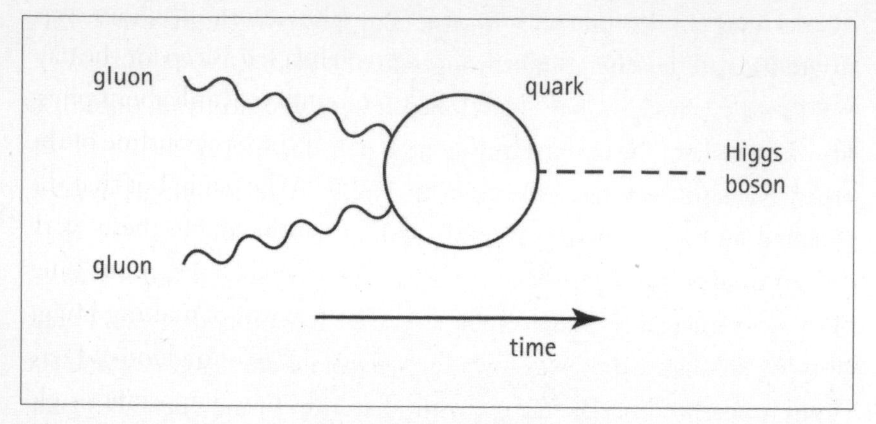

A Feynman diagram representing two gluons fusing together to create a Higgs boson, via the intermediate step of virtual quarks.

computed by following a series of straightforward rules. These rules can be confusing at first glance; for example, a particle going backward in time counts as an antiparticle, and vice versa. When two particles join to make a third (or one decays into two), the total energy and all other conserved quantities must balance. But the virtual particles—the ones that move around in the interior of the diagram but aren't present in the initial collection or the final products—don't have to have the same mass that a real particle would have. The right way to think about the diagram above is that two vibrations in the gluon field come together and set up a vibration in the quark field, which ultimately produces a vibration in the Higgs field. What we actually see are two gluon particles joining to create a Higgs boson.

The first person to realize that "gluon fusion" was a promising way to create Higgs bosons was Frank Wilczek, the American theorist who had helped pioneer our understanding of the strong interactions—work he did in 1973 as a graduate student, and for which he eventually shared the Nobel Prize. In 1977, he was on the faculty at Princeton, but he took time to visit Fermilab over the summer. Even the world's great thinkers must take care of the mundane challenges of everyday life, and on this occasion Wilczek had spent a long day attending to his wife,

Betsy Devine, and their infant daughter, Amity, both of whom were struggling with illness. After his wife and daughter fell asleep for the day, Wilczek took a walk around the Fermilab grounds to think about physics. Even at that time it was becoming clear that the basic outline of the Standard Model was "pretty much a done deal," as he put it, but that the properties of the Higgs boson were relatively unexplored. His thesis work had given him a great fondness for gluons and their interactions, and while walking he realized that gluons were a great way of making Higgs bosons (and that Higgs bosons could in turn decay into gluons). Here we are thirty-five years later, and this process is the most important single way that the Higgs is produced at the LHC. On the same walk, Wilczek also came up with the idea for the "axion," a hypothetical low-mass cousin of the Higgs that is now a promising candidate for constituting the dark matter in the universe. A testament to the importance of long, peaceful walks to the progress of physics.

In Appendix Three, we discuss the various ways that particles can interact in the Standard Model, and the Feynman diagrams corresponding to each possibility. Not carefully enough to get anyone a PhD in physics, but hopefully enough to give you the general idea. One thing should be clear: It's a bit of a mess. It's easy to say, "We smash protons together and wait for a Higgs to come out," but it's a lot of work to sit down and do the calculations carefully. When all is said and done, a number of different processes contribute to creating Higgs bosons at the LHC: the fusion of two gluons as we just discussed; the analogous fusion of a W^+ with a W^-, or two Z bosons, or a quark and an antiquark; and the production of a W or Z that spits off a Higgs before going on its way. Details depend on the mass of the Higgs, as well as the energy of the original collisions. Calculating the relevant processes provides full employment for theoretical physicists.

What comes out

So you've made a Higgs boson! Congratulations. Now comes the tricky part: How are you ever going to know?

Heavy particles tend to decay, and the Higgs is very heavy indeed. The lifetime of the Higgs is estimated to be somewhat less than a zeptosecond (10^{-21} seconds), which means it gets to travel less than a billionth of an inch between when it's produced and when it decays. Even with the very advanced detectors inside ATLAS and CMS, there's no way we're seeing that. Instead, we see what the Higgs decays into. We will also see things that other non-Higgs particles decay into, many of which look just like what the Higgs decays into. The trick is to pick out the tiny signal from the huge amount of background noise.

The first step is to figure out exactly what your Higgs is going to decay into, and how often. In general, the Higgs likes to couple to heavy particles, so we might expect it to decay frequently to top and bottom quarks, W and Z bosons, and the tau lepton; not so much to lighter particles like up and down quarks or electrons. And that's basically right, although there are subtleties (as you knew there would be).

For one thing, the Higgs can't decay into something that's heavier than it is. It can temporarily convert into heavy virtual particles that themselves quickly decay away, but processes like that become very rare if the virtual particles are much heavier than the original Higgs. If the Higgs were 400 GeV, it could readily decay into a top quark and an antitop, which come in at 172 GeV each. But for a more realistic Higgs mass like 125 GeV, top quarks are unavailable, and bottom quarks are the favored decay mode. That's one reason why heavier versions of the Higgs (up to 600 GeV) would have actually been easier to find, even though it takes more energy to create them—the rate of decay into heavy particles is much higher.

The figure shows a pie chart giving the approximate ratio of different decay modes for a Higgs boson with a mass of 125 GeV, according

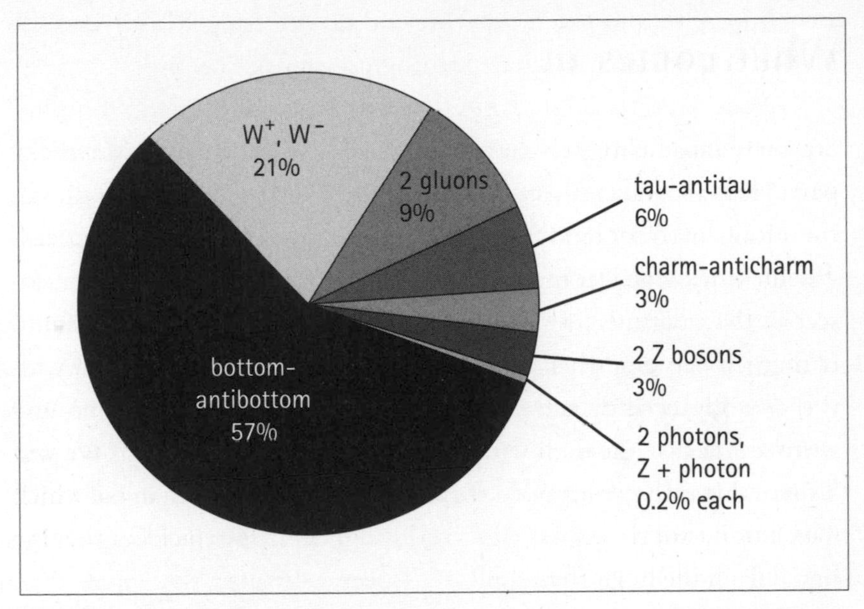

Probability of a Higgs boson with mass 125 GeV decaying into different particles. Numbers don't add up to exactly 100 percent due to rounding.

to the Standard Model. The Higgs will decay into a bottom quark and an antibottom most of the time, but there are a number of other important possibilities. Although this value for the Higgs mass makes it hard to detect, once we do there's a tremendous amount of interesting physics to be studied—we can measure each decay mode separately and compare it with the predictions. Any deviation would be a sign of physics beyond the Standard Model, such as additional particles or unusual interactions. We've even seen hints that such deviations might actually exist.

We're nowhere near done yet, however. Hearken back to our discussion of particle detectors in Chapter Six, where we saw how different layers of the experimental onion helped us pinpoint different outgoing particles: electrons, photons, muons, and hadrons. Then look back at this pie chart. More than 99 percent of the time, the Higgs decays into something that we don't observe directly in our detector. Rather, the Higgs decays into something that then decays (or transforms) into something else, and *that* is what we end up detecting. This makes life harder, or more interesting, depending on your perspective.

About 70 percent of the time, the Higgs decays into quarks (bottom-antibottom or charm-anticharm) or gluons. These are colored particles, which aren't found by themselves in the wild. When they are produced, the strong interactions kick in and create a cloud of quarks/antiquarks/gluons, which congeal into jets of hadrons. Those jets are what we detect in the calorimeters. The problem—and it's a very big problem—is that jets are produced by all sorts of processes. Smash protons together at high energy and you'll be making jets by the bushelful, and a tiny fraction of the total will be the result of decaying Higgs bosons. Experimentalists certainly do their best to fit this kind of signal to the data, but it's not the easiest way to go about detecting the Higgs. In the first full year's run of the LHC, it's been estimated that more than 100,000 Higgs bosons were produced, but most of them decayed into jets that were lost in the cacophony of the strong interactions.

When the Higgs doesn't decay directly into quarks or gluons, it usually decays into W bosons, Z bosons, or tau-antitau pairs. All of these are useful channels to look at, but the details depend on what these massive particles decay into themselves. When tau pairs are produced, they generally decay into a W boson of the appropriate charge plus a tau neutrino, so the analysis is somewhat similar to what happens when the Higgs decays into Ws directly. Often, the decaying W or Z will produce quarks, which lead to jets, which are hard to pick out from above the background. Not impossible—hadronic decays are looked at very seriously by the experimenters. But it's not a clean result.

Some of the time, however, the W and Z bosons can decay purely into leptons. The W can decay into a charged lepton (electron or muon) and its associated neutrino, while the Z can decay directly into a charged lepton and its antiparticle. Without jets getting in the way, these signals are relatively clean, although quite rare. The Higgs decays to two charged leptons and two neutrinos about 1 percent of the time, and into four charged leptons about 0.01 percent of the time. When the W

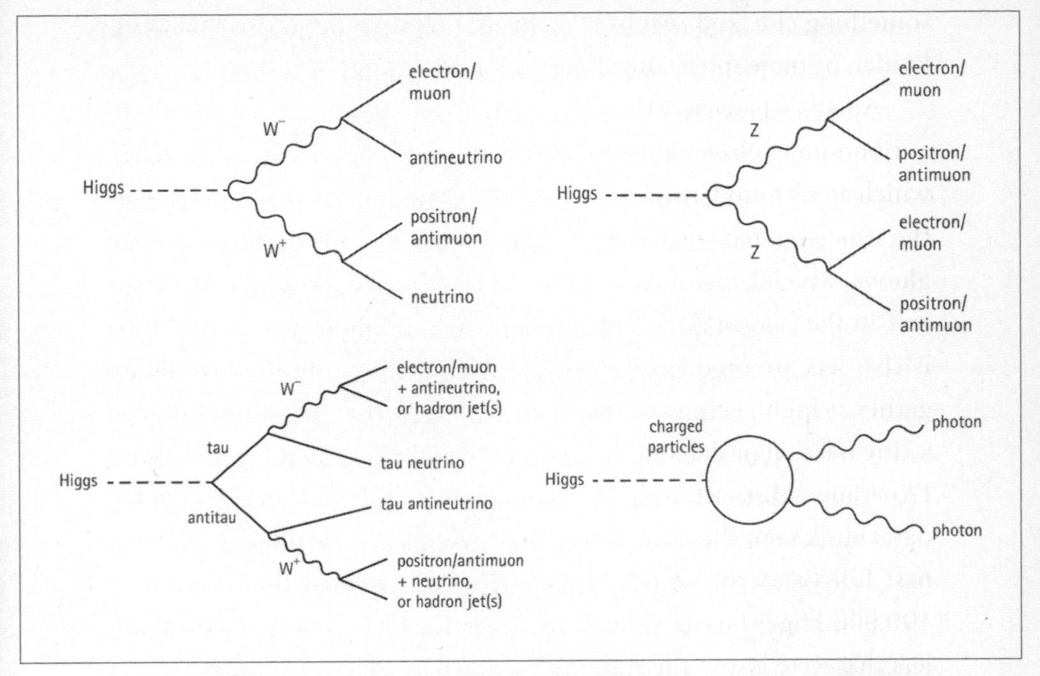

Four promising decay modes for discovering a Higgs boson at 125 GeV. The Higgs can decay into two W bosons, which then (sometimes) decay into electrons or muons and their neutrinos. Or it can decay into two Z bosons, which then (sometimes) decay into electrons or muons and their antiparticles. Or it can decay into a tau-antitau pair, which then decays into neutrinos and other fermions. Or it can decay into some charged particle that then converts into two photons. These are all rare processes but relatively easy to pick out at LHC experiments.

decays create neutrinos, the missing energy makes these events hard to pin down, but they're still useful. The four-charged-lepton events from Z decays have no missing energy to confuse things, so they are absolutely golden, though so uncommon that they're very hard to find.

And sometimes, through a bit of help from virtual particles with electric charge, the Higgs can decay into two photons. Because photons are massless they don't couple directly to the Higgs, but the Higgs can first create a charged massive particle, and then that can transform into a pair of photons. This happens only about 0.2 percent of the time, but it ends up being the clearest signal we have for a Higgs near 125 GeV. The rate is just large enough that we can get a sufficient number of events, and the background is small enough that it's possible to see the Higgs signal sticking out above the background. The best evidence we've gathered for the Higgs has come from two-photon events.

This whirlwind tour of the different ways a Higgs can decay is just a superficial overview of the tremendous amount of theoretical effort that has gone into understanding the properties of the Higgs boson. That project was launched in 1975 in a classic paper by John Ellis, Mary K. Gaillard, and Dimitri Nanopoulos, all of whom were working at CERN at the time. They investigated how one could produce Higgs bosons, as well as how to detect them. Since then a large number of works have reconsidered the subject, including an entire book called *The Higgs Hunter's Guide*, by John Gunion, Howard Haber, Gordon Kane, and Sally Dawson, which has occupied a prominent place on the bookshelves of a generation of particle physicists.

In the early days, there was much we hadn't figured out about the Higgs. Its mass was always a completely arbitrary number, which we have only homed in on through diligent experimental efforts. In their paper Ellis, Gaillard, and Nanopoulos gave a great deal of attention to masses of 10 GeV or less; had that been right, we would have found the Higgs ages ago, but nature was not so kind. And they couldn't resist closing their paper with "an apology and a caution":

> We apologize to experimentalists for having no idea what is the mass of the Higgs boson . . . and for not being sure of its couplings to other particles, except that they are probably all very small. For these reasons, we do not want to encourage big experimental searches for the Higgs boson, but we do feel that people doing experiments vulnerable to the Higgs boson should know how it may turn up.

Fortunately, big experimental searches were eventually encouraged, although it took some time. And now they are paying off.

Achieving significance

Searching for the Higgs boson has frequently been compared to looking for a needle in a haystack, or for a needle in a large collection of many haystacks. David Britton, a Glasgow physicist, who has helped put together the LHC computing grid in the United Kingdom, has a better analogy: "It's like looking for a bit of *hay* in a haystack. The difference being that if you look for a needle in a haystack you know the needle when you find it, it's different from all the hay . . . the only way to do it is to take every bit of hay in that haystack, line them all up, and suddenly you'll find there's a whole bunch at one particular length, and this is exactly what we're doing."

That's the challenge: Any individual decay of the Higgs boson, even into "nice" particles like two photons or four leptons, can also be produced (and will be, more often) by other processes that have nothing to do with the Higgs. You're not simply looking for a particular *kind* of event, you're looking for a slightly larger *number* of events of a certain kind. It's like you have a haystack with stalks of hay in all different sizes, and what you're looking for is a slight excess of stalks at one particular size. This is not going to be a matter of examining the individual bits of hay closely; you're going to have to turn to statistics.

To wrap our heads around how statistics will help us, let's start with a much simpler task. You have a coin, which you can flip to get heads or tails, and you want to figure out whether the coin is "fair"—i.e., if it comes up heads or tails with exactly fifty-fifty probability. That's not a judgment you can possibly make by flipping the coin just two or three times—with so few trials, no possible outcome would be truly surprising. The more flips you do, the more accurate your understanding of the coin's fairness is going to be.

So you start with a "null hypothesis," which is a fancy way of saying "what you expect if nothing funny is going on." For the coin, the null hypothesis is that each flip has a fifty-fifty chance of giving heads or tails. For the Higgs boson, the null hypothesis is that all of your data is produced as if there is *no* Higgs. Then we ask whether the actual data are consistent with the null hypothesis—whether there's a reasonable chance we would have obtained these results with a fair coin, or with no Higgs lurking there.

Imagine that we flip the coin one hundred times. (Really we should do a lot more than that, but we're feeling lazy.) If the coin were perfectly fair, we would expect to get fifty heads and fifty tails, or something close to that. We wouldn't be surprised to get, for example, fifty-two heads and forty-eight tails, but if we obtained ninety-three heads and only seven tails we'd be extremely suspicious. What we'd like to do is quantify exactly how suspicious we should be. In other words, how much deviation from the predicted fifty-fifty split would we need to conclude that we weren't dealing with a fair coin?

There's no hard and fast answer to this question. We could flip the coin a billion times and get heads every time, and in principle it's possible that we were just really, really lucky. That's how science works. We don't "prove" results like we can in mathematics or logic; we simply add to their plausibility by accumulating more and more evidence. Once the data are sufficiently different from what we would expect under the null hypothesis, we reject it and move on with our lives, even if we haven't attained metaphysical certitude.

Because we're considering processes that are inherently probabilistic, and we look only at a finite number of events, it's not surprising to get some deviation from the ideal result. We can actually calculate how much deviation we would typically expect, which is labeled with the Greek letter sigma, written as σ. This lets us conveniently express how big an observed deviation actually is—how much bigger is it than sigma? If the difference between the observed measurement and the ideal prediction is twice as big as the typical expected uncertainty, we say we have a "two-sigma result."

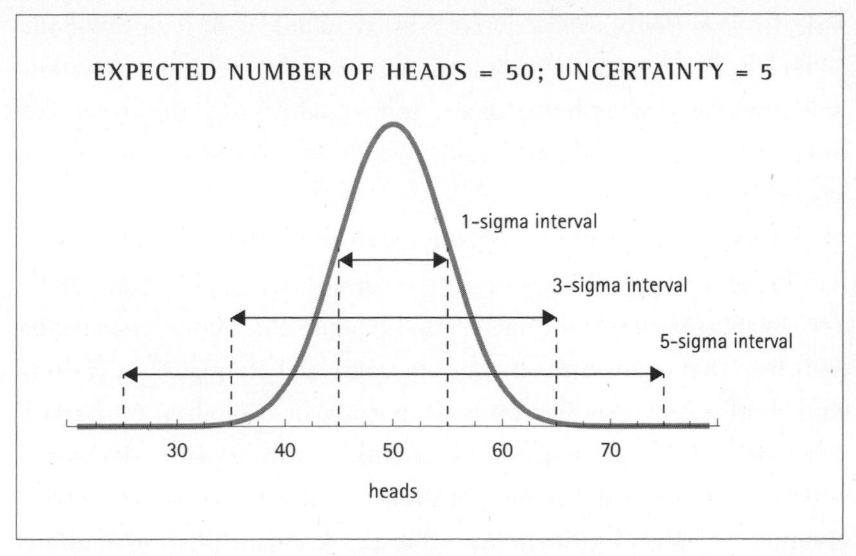

Confidence intervals for flipping a coin 100 times, which has an expected value of 50 and an uncertainty of sigma = 5. The one-sigma interval stretches from 45 to 55, the three-sigma interval stretches from 35 to 65, and the five-sigma interval stretches from 25 to 75.

When we make a measurement, the variability in the predicted outcome often takes the form of a bell curve, as shown in the figure above. Here we are graphing the likelihood of obtaining different outcomes (in this example, the number of heads when we flip a coin a hundred times). The curve peaks at the most likely value, which in this case is fifty, but there is some natural spread around that value. This spread, the width of the bell curve, is the uncertainty in the prediction, or equivalently the value of sigma. For flipping a fair coin a hundred times, sigma = 5, so we would say, "We expect to get heads fifty times, plus or minus five."

The nice thing about quoting sigma is that it translates into the probability that the actual result would be obtained (even though the explicit formula is a mess, and usually you just look it up). If you flip a coin one hundred times and get between forty-five and fifty-five heads, we say you are "within one sigma," which happens 68 percent of the time. In other words, a deviation of more than one sigma happens about 32 percent of the time—which is quite often, so a one-sigma deviation isn't anything to write home about. You wouldn't judge the coin to be unfair just because you got fifty-five heads and forty-five tails in one hundred flips.

Greater sigmas correspond to increasingly unlikely results (if the null hypothesis is right). If you got sixty heads out of a hundred, that's a two-sigma deviation, and such things happen only about 5 percent of the time. That seems unlikely but not completely implausible. It's not enough to reject the null hypothesis, but maybe enough to raise some suspicion. Getting sixty-five heads would be a three-sigma deviation, which occurs about 0.3 percent of the time. That's getting pretty rare, and now we have a legitimate reason to think something fishy is going on. If we had gotten seventy-five heads out of one hundred flips, that would be a five-sigma result, something that happens less than one in a million times. We are therefore justified in concluding that this was not just a statistical fluke, and the null hypothesis is not correct—the coin is not fair.

Signal and background

Particle physics, since it is powered by quantum mechanics, is a lot like coin flipping: The best we can do is predict probabilities. At the LHC, we smash protons together and predict the probability of different interactions occurring. For the particular case of the Higgs search, we consider different "channels," each of which is specified by the particles that are captured by the detector: There's the two-photon channel, the two-lepton channel, the four-lepton channel, the two-jets-plus-two-leptons channel, and so on. In each case, we add up the total energy of the outgoing particles, and the machinery of quantum field theory (aided by actual measurements) allows us to predict how many events we expect to see at every energy, typically forming a smooth curve.

That's the null hypothesis—what we expect without any Higgs boson. If there is a Higgs at some specific mass, its main effect is to give a boost to the number of events we expect at the corresponding energy: A 125 GeV mass Higgs creates some extra particles with a total energy of 125 GeV, and so on. Creating a Higgs and letting it decay provides a mechanism (in addition to all the non-Higgs processes) to produce particles that typically have the same total energy as the Higgs mass, leading to a few additional events over the background. So we go "bump hunting"—is there a noticeable deviation from the smooth curve we would see if the Higgs wasn't there?

Predicting what the expected background is supposed to be is by no means an easy task. We know the Standard Model, of course, but just because we know what the theory is doesn't mean it's easy to make a prediction. (The Standard Model also describes the earth's atmosphere, but it's not easy to predict the weather.) Powerful computer programs do their best to simulate the most likely outcomes of the proton collisions, and those results are run through a simulation of the detectors themselves. Even so, we readily admit that some rates are easier to measure than to predict. So it is often best to do a "blind" analysis—use some method to disguise the actual data of interest, by adding fake data to it or simply not looking at certain events, then making every effort to understand the boring data in other regions, and only once the best possible understanding is achieved do we "open the box" and look at the data where our particle might be lurking. A procedure like this helps to ensure that we don't see things just because we want to see them; we only see them when they're really there.

It wasn't always so. In his book *Nobel Dreams*, journalist Gary Taubes tells the story of Carlo Rubbia's work in the early 1980s that discovered the W and Z bosons and won him a Nobel Prize, as well as his less successful attempts to win a second Nobel by finding physics beyond the Standard Model. One of the tools that Rubbia's team used in their analysis was the Megatek, a computer system that could display the data from particle collisions and let the user rotate the view in three dimensions by operating a joystick. Rubbia's lieutenants, American James Rohlf and Englishman Steve Geer, became masters of the Megatek. They were able to glance at an event, twirl it a bit, pick out the important particle tracks, and declare with confidence that they were seeing a W or a Z or a tau. "You have all this computing," in Rubbia's words, "but the purpose of all this tremendous data analysis, the one fundamental bottom line, is to be able to let the human being give the final answer. It's James Rohlf looking at the f***ing event who will decide whether this is a Z or not." No longer—we have a lot more data now, but the only way to really understand what you're seeing is to hand it over to the computer.

Whenever there is some excitement about a purported experimental result, your first instinct should be to ask, "How many sigma?" Within particle physics, an informal standard has arisen over the years, according to which a three-sigma deviation is considered "evidence for" something going on, while a five-sigma deviation is needed to claim "discovery of" that something. That might seem unduly demanding, since a three-sigma result is already something that only happens 0.3 percent of the time. But the right way to think about it is, if you look at three hundred different measurements, one of them is likely to be a three-sigma anomaly just by chance. So sticking to five sigma is a good idea.

At the December 2011 seminars, the peak near 125 GeV had a significance of 3.6 sigma in the ATLAS data, and 2.6 sigma in the CMS data (which are completely independent). Suggestive, but not enough to claim a discovery. Speaking against the significance of the result was the "look-elsewhere effect"; the simple fact that, as we just alluded to, large deviations are likely if you look at many different possible measurements, which the two LHC experiments were certainly doing. But at the same time, the fact that the two experiments saw bumps in the same place was extremely suggestive. Taken all together, the sense of the community was that the experiments probably were on the right track, and we probably were seeing the first glimpses of the Higgs—but only more data would tell for sure.

When the predictions you are testing involve probabilities, the importance of collecting more data cannot be overemphasized. Think back to our coin-flipping example. If we had only flipped the coin five times instead of a hundred, the biggest possible deviation from the expected value would have been to get all heads (or all tails). But the chance of that happening is more than 6 percent. So even for a completely unfair coin, we wouldn't be able to claim as much as a two-sigma deviation from fairness. On *Cosmic Variance*, a group blog I contribute to that is hosted by *Discover* magazine, I put up a post on the day before the CERN seminars, entitled "Not Being Announced Tomorrow: Discovery of the Higgs Boson." It's not that I had any inside information; it's just that we all knew how much data the LHC had produced up to that time, and it simply wasn't enough to claim a five-sigma discovery of the Higgs. That would have to wait for more data.

The bear is caught

The general feeling among physicists was that if the 2011 hints were signs of something real, the data collected in 2012 would be enough to reach the magical five-sigma threshold necessary to declare a discovery. We knew how many collisions were happening at the LHC, and the feeling worldwide was that we would be able to declare discovery (or crushing disappointment) a year later, in December 2012.

After its yearly winter shutdown, the LHC resumed collecting data in February. The International Conference on High Energy Physics (ICHEP) in Melbourne was planned for early July, and both experiments anticipated giving updates of their progress at that meeting. Conditions in 2012 were somewhat different from those in 2011, so it wasn't immediately obvious how quickly progress could be made. They were running at a higher energy—8 TeV rather than 7 TeV—and also at a higher luminosity, so they were getting more events per second.

Both of those sound like improvements, which they are, but they are also challenges. Higher energy means slightly different interaction rates, which means slightly different numbers of background events, which means you have to calibrate the new data separately from the old data. Higher luminosity means more collisions, but many of those collisions are happening simultaneously in the detector. This leads to "pileup"—you see a bunch of particle tracks but have to work hard to separate which ones came from which collisions. It's a nice problem to have; but it's still a problem you have to solve, and that takes time.

The ICHEP is a major international event, and a logical venue at which to provide an update on the progress of the Higgs search after the new data had started coming in at higher energies. What people expected to hear was that the machine was doing great, and ideally that the statistical significance of the December hints was growing rather than shrinking. The LHC was scheduled to pause in its data collection in early June for routine maintenance purposes, and that was chosen as a natural point at which to look at the data carefully and see what they had.

Both experiments were analyzing their data blind. The "box" containing the true data in the region of interest was opened on June 15, leaving about three weeks for the experimentalists to figure out what they had and how to present it in Melbourne.

Almost immediately the rumors started flying. They were a little bit more vague than they had been in December, which is understandable; the experimenters themselves were scrambling to figure out what it was that they had. In the end, I don't know of any rumors that got the final result precisely correct. But the general tenor was unmistakable: They were seeing something big.

What they were seeing, of course, was a new particle—the Higgs, or something near enough. Even a glance at the data was enough to see that. The stakes were immediately raised; a simple update wasn't going to be an appropriate tack to take when the results were presented to the public. You either have a discovery, or you don't; and if you do, you don't bury the lede, you trumpet it to the world.

As subgroups within the experiments frantically analyzed the data in the various different channels, higher-ups debated how best to deploy the trumpets. On the one hand, both experiments were scheduled to give updates in Melbourne, and it would seem petty to pull out. On the other, there were hundreds of physicists at CERN who weren't going to fly around the world, and this day belonged to them as much as to anyone. In the end, a compromise was reached: Each experiment would give a seminar on the day the conference opened, but the seminars would be located in Geneva and simulcast in Australia.

If that weren't enough to convince people on the outside that important news was coming, word quickly spread that CERN was inviting big names to be present at the seminars. Peter Higgs, now age eighty-three, was at a summer school in Sicily at the time; he was scheduled to fly back to Edinburgh, his travel insurance had run out, and he didn't have any Swiss francs with him. But he changed his plans after John Ellis, the eminent theorist at CERN and longtime Higgs boson aficionado, left him a phone message: "Tell Peter that if he doesn't come to CERN on Wednesday, he will very probably regret it." He came, as did François Englert, Gerald Guralnik, and Carl Hagen, other theorists who had helped pioneer the Higgs mechanism.

In December 2011, I was back in California and slept right through the seminars, which started at five a.m. Pacific time. But in July 2012, I managed to book a flight to Geneva and was there at CERN for the big day. I and many others were running from building to building at the lab, scrambling to get the proper credentials. At one point I had to sweet-talk my way past a security guard to get back into a building from which I had just exited, and explained that I was kind of short on time. "Why is everybody in a hurry today?" he asked.

As in December, hundreds of people (mostly younger folks) had camped out overnight to get good seats in the auditorium. Gianotti once again gave the talk reporting results from ATLAS, but Tonelli's term as CMS spokesperson had run out and the CMS talk was given by his successor, Joe Incandela, from the University of California, Santa Barbara. Incandela and Gianotti had both cut their teeth working together on UA2, one of the detectors at CERN's previous hadron collider, and they had searched for Higgs bosons in the data from that experiment. Now they were about to see their long-standing quest come to fruition.

Everyone in the room knew that all this fuss wouldn't be happening if the signal had gone away. The primary question was, how many sigma? Between rumors and back-of-the-envelope estimations, the prevailing opinion seemed to favor the idea that each experiment would reach four-sigma significance, but not quite five. Combining the two, however, might bump us over the five-sigma threshold. But combining data from two different experiments is much trickier than it sounds, and it didn't seem feasible that it could have been done over just the past three weeks. There was more than a little worry that we were going to be tantalized once more, but not quite able to claim a discovery.

We needn't have worried. Incandela, who spoke first, went through the different channels that had been analyzed by CMS one by one. Two-photon events came first, and they displayed a noticeable peak just where we were hoping, at 125 GeV. The significance was 4.1 sigma—more than in the previous year, but not a discovery. Then came events with four charged leptons, which result from the Higgs decaying into two Z bosons. Another peak, in the same place, this time with 3.2 sigma significance. On his sixty-fourth PowerPoint slide, Incandela revealed what you get when you combine these two channels together: 5.0 sigma. The wait was over. We found it.

Gianotti, like Incandela, went out of her way to praise the hard work of everyone who helped keep the LHC running, and she emphasized the care the ATLAS collaboration went through to analyze their data. When she turned to the two-photon results, there was once again an evident peak at 125 GeV. This time the significance was 4.5 sigma. The four-lepton results also fell into line: a tiny peak, but discernible, with a significance of 3.4 sigma. Combining them gave an overall significance of exactly 5.0 sigma. At the end of her talk, Gianotti thanked nature for putting the Higgs where the LHC could find it.

ATLAS found a Higgs mass of 126.5 GeV, while CMS got 125.3 GeV, but the measurements are within the expected uncertainty of each other. CMS analyzed more channels in addition to two photons and four leptons, and as a result their final significance ended up dropping just a tiny amount, to 4.9 sigma. But again, that's consistent with the overall picture. The agreement between the two experiments was amazing, and crucially important. If the LHC had only one detector looking for the Higgs, the physics community would be much more hesitant to take the results at face value. As it was, hesitancy was thrown to the wind. This was a discovery.

After the seminars were over, Peter Higgs became emotional. He later explained, "During the talks I was still distancing myself from it all, but when the seminar ended, it was like being at a football match when the home team had won. There was a standing ovation for the people who gave the presentation, cheers and stamping. It was like being knocked over by a wave." In the pressroom afterward, reporters tried to get more comments from him, but he demurred, saying that the focus on a day like this should be on the experimenters.

In retrospect, a lot of things went right in the first half of 2012 to enable a Higgs discovery earlier than most people expected. The LHC was going full steam, collecting more events in just a few months of operation than it had in all of 2011. Pileup was a challenge, but the data analysts met it heroically, and the overwhelming fraction of events were successfully reconstructed. The higher energy pushed up the rate at which Higgs bosons were produced. And the teams had honed their analysis routines, managing to squeeze more significance out of their data than before. All these improvements ended up giving particle physicists Christmas in July.

What is it?

After the seminars were over, Incandela was reflective. "You often think that, once you've discovered something, it's an end. What I've learned in science is that it's almost always a beginning. There's almost always something very big, just right there, that is within reach, and you just have to go for it. So you can't let down your guard!"

There is no question that CMS and ATLAS have found a new particle. There is very little question that the new particle resembles the Higgs boson; its decay rates into different channels match up roughly with what the Standard Model Higgs is expected to do if its mass is 125 GeV or so. But there's plenty of reason to wonder whether it really is the simplest Higgs, or something more subtle. There are tiny hints in the data that may indicate that this new particle is not just the minimal Higgs. It's far too early to tell whether those hints are real; they could easily go away, but we can rest assured that the experiments will be following up on them to figure out what's really going on.

Remember that particles don't appear in the detector with labels. When we say that we've found something consistent with a Higgs boson, we're referring to the fact that the Standard Model makes very specific predictions once the mass of the Higgs is fixed. There are no other free parameters; knowing that one number allows us to say precisely how many decays there will be into each channel. Saying that we see something like the Higgs is saying that we see the right amount of excess events in all the channels where they should be visible, not just in one.

The figures included in the color insert show the data from ATLAS and CMS in 2011 and early 2012, looking specifically at collisions that created two photons. What we see are the numbers of events in which the two photons total up to a specific energy. Notice how few of these events there actually are. The experiment sees hundreds of millions of interactions per second, of which a couple hundred per second pass through the trigger and are recorded for posterity; but in a year's worth of data, we get only a thousand or so events at each energy.

The dashed curve in the figure is the prediction for the background—what you would expect without a Higgs. The solid line is what happens when we include the ordinary Standard Model Higgs, with a mass of 125 GeV. Both curves show a small bump with a couple hundred more events than expected. You can't say which events are Higgs decays, and which decays are background, but you can ask whether there is a statistically significant excess. There is.

Closer inspection reveals something funny about these data. One of the reasons we were surprised to find the Higgs so quickly in 2012 is that the experiments actually observed more events than they should have. The significance of the two-photon bump in the ATLAS data is 4.5 sigma, but with the number of collisions analyzed the Standard Model predicts that we should have reached only 2.4 sigma. Likewise, in CMS, the significance was 4.1 sigma, but it was expected to reach only 2.6 sigma.

In other words, there were more excess events with two photons than we should have seen. Not too many more; the sizes of the bumps are a bit bigger than expected but still within the known uncertainties. But the fact that they are consistent between both experiments (and consistent with ATLAS's result from 2011 alone) is intriguing. There is no question we will need more data to see whether this discrepancy is real or just a tease.

The CMS data presented another small but noticeable puzzle. While ATLAS stuck with the robust channels of two photons or four charged leptons, CMS also analyzed three noisier channels: tau-antitau, bottom-antibottom, and two Ws. As might be expected, the bottom-antibottom and WW channels didn't give statistically significant results (although more data will certainly improve the situation). The tau-antitau analysis, however, was a puzzle: No excess was seen at 125 GeV, even though the Standard Model predicts that it should be. This was not quite a statistically significant discrepancy, but it's interesting. Indeed, the slight tension with the tau data is what brought the final significance of the full CMS analysis down to 4.9 sigma, even though the two-photon and four-lepton channels alone had achieved five sigma.

What could be going on? None of these hints is serious enough to be sure that anything at all is going on, so it might not be worth taking the discrepancies too seriously. But as theorists, that's what we do for a living. Within a day or two after the seminars, theory papers were already appearing online, attempting to sort it all out.

It's easy to give one simple example of the kind of thing that people are thinking about. Remember how the Higgs decays into two photons. Because photons are massless, and therefore don't couple directly to the Higgs, the only way this can happen is via some intermediate virtual particle that is both massive (so it couples to the Higgs) and electrically charged (so that it couples to photons).

By the rules of Feynman diagrams, we are instructed to calculate the rate for this process by adding up independent contributions from all the different massive charged particles that could appear in the loop inside this diagram. We know what the Standard Model particles are, so that's not hard to do. But new particles can easily change the answer by contributing to those virtual processes, even if we've not yet been able to detect them directly. So the anomalously large number of events might be the first signal of particles beyond the Standard Model, helping the Higgs decay into two photons.

Details matter, of course; if the new particles you have in mind also change the rates of other measured processes, you might be in trouble. But it's exciting to think that by studying the Higgs we might be learning not only about that particle itself but also about other particles yet to be found.

Don't let down your guard.

TEN
SPREADING THE WORD

In which we draw back the curtain on the process by which results are obtained and discoveries are communicated.

With all the solemn British rectitude he could summon, correspondent John Oliver was putting tough questions to Walter Wagner, the man who had gone to court to stop the Large Hadron Collider from beginning operations. A serious charge had been leveled: the LHC was a hazard to the very existence of life on earth.

JO: So, roughly speaking, what are the chances the world is going to be destroyed? Is it one in a million, one in a billion?

WW: Well, the best we can say right now is about a one-in-two chance.

JO: Hold on a second. It's . . . fifty-fifty?

WW: Yeah, fifty-fifty . . . If you have something that can happen, and something that won't necessarily happen, it's going to either happen, or it's going to not happen, and, so, the best guess is one in two.

JO: I'm not sure that's how probability works, Walter.

As the LHC was starting up in 2008, physicists tried their best to spread the word that this was a machine that would help us find the Higgs boson, perhaps reveal supersymmetry for the first time, and possibly discover exciting and exotic phenomena such as dark matter or extra dimensions. But against this uplifting story of human curiosity triumphant, a countervailing narrative struggled for people's attention: The LHC was a potentially dangerous experiment that would re-create the Big Bang and potentially destroy the world.

At the time, the mad-scientists-out-of-control scenario was winning the competition for attention. It's not that journalists were willing to ignore the truth and seek out sensationalism for its own sake. (At least, not most of them. In the United Kingdom, the *Daily Mail* tabloid ran a big headline, ARE WE ALL GOING TO DIE NEXT WEDNESDAY?) Rather, much like the label "God Particle," the disaster scenarios seemed to be a mandatory part of any news story. Once the idea is posed that *just maybe* the LHC could kill everyone on earth—even if it was something of a long shot—that's the question that people wanted to see addressed. Added to the mix was Walter Wagner, a litigious former nuclear safety officer, who brought a quixotic suit against the LHC in Hawaii. After the case was thrown out of court on (fairly evident) jurisdictional grounds, Wagner appealed to federal court. A three-judge panel finally dismissed the case in 2010, with a pithy conclusion:

> Accordingly, the alleged injury, destruction of the earth, is in no way attributable to the U.S. government's failure to draft an environmental impact statement.

CERN and other physics organizations took the need to proceed safely very seriously, sponsoring multiple expert reports on the subject, all of which concluded that the risk of disaster was completely negligible. Oliver's interview, which allowed Wagner to discredit himself with his own words, was one of the very few news reports to take an appropriate angle on the topic. It appeared on Jon Stewart's *The Daily Show,* a satirical news program from Comedy Central channel. Only a comedy program was smart enough to treat the LHC disaster worry as the farce that it was.

One thing working against the physicists was their natural inclination to be both precise and honest, often to the detriment of getting their point across. The fears that the LHC could destroy the world were based in part on respectable, if speculative, physical theories. If gravity is much stronger than usual at the high energies of an LHC particle collision, for example, it's possible to make tiny black holes. Everything we know about physics predicts that such a black hole will evaporate harmlessly away. But it's possible that everything we know is wrong. So maybe black holes are formed and are stable, and the LHC will produce them, and they will settle into the earth's core and gradually eat at it from the inside, leading to a collapse of the planet over the course of time. You can calculate how much time it would actually take, and the answer turns out to be much longer than the current age of the universe. Of course, your calculations could be incorrect. But in that case, collisions of high-energy cosmic rays should be producing tiny black holes all over the universe. (The LHC isn't doing anything the universe doesn't do at much higher energies all the time.) And those black holes should eat up white dwarfs and neutron stars, but we see plenty of white dwarfs and neutron stars in the sky, so that can't be quite right either.

You get the point. There are many variations on the theme, but the general pattern is universal: We can come up with very speculative scenarios that seem dangerous, but upon closer inspection the most dangerous possibilities are already ruled out by other considerations. But because scientists like to be precise and consider many different possibilities, they tend to dwell lovingly on all the scary-sounding scenarios before reassuring us that they are all quite unlikely. Every time they should have said, "No!" they tended to say, "Probably not, the chance is really very small," which doesn't have the same impact. (A shining counterexample is CERN theorist John Ellis, who was asked by *The Daily Show* what chance there was that the LHC would destroy the earth, and simply replied, "Zero.")

Imagine opening your refrigerator and reaching for a jar of tomato sauce, planning to make pasta for tonight's dinner. An alarmist friend grabs you before you can open the lid, saying, "Wait! Are you sure that opening that jar won't release a mutant pathogen that will quickly spread and wipe out all life on earth?" The truth is, you can't be sure, with precisely 100 percent certainty. There are all sorts of preposterously small probability disaster scenarios that we ignore in our everyday lives. It's conceivable that turning on the LHC will start a chain of events that destroys the earth, but many things are conceivable; what matters is whether they are reasonable, and in this case none of them was.

Fighting against the doomsayers turned out to be good practice for the physics community. The level of public scrutiny given to the search for the Higgs boson is unprecedented. Scientists, who are at their best when discussing abstract and highly technical ideas with other scientists, have had to learn to craft a clear and compelling message for the outside world. In the long run, that can only be good news for science.

Making the sausage

One of the biggest misconceptions many people have about results that come from giant particle physics experiments is about the journey from taking data to announcing a result. It's not an easy one. In science, the traditional way that results are communicated and made official is through papers published in peer-review journals. That's certainly true for ATLAS and CMS, but the complexity of the experiments guarantees that essentially the only competent referees are the collaboration members themselves. To deal with this state of affairs, each experiment has set up an extremely rigorous and demanding procedure that must be carried out before new results can be shared with the public.

The thousands of collaborators on the LHC experiments are mostly not employed by CERN. A typical working physicist will be a student, professor, or postdoc (a research position in between the PhD and a faculty job) at a university or laboratory somewhere in the world, although they may spend a substantial portion of their year in Geneva. Most often, the first step toward a publishable paper is that one of these physicists asks a question. It might be a perfectly obvious question: "Is there a Higgs boson?" Or it could be something more speculative: "Is electric charge really conserved?" "Are there more than three generations of fermions?" "Do high-energy particle collisions create miniature black holes?" "Are there extra dimensions of space?" Questions may be inspired by a new theoretical proposal, or an unexplained feature of some existing data, or simply by the new capabilities of the machine itself. Experimentalists are generally down-to-earth people, at least in their capacity as working scientists, so they tend to ask questions that can be addressed by the flood of data the LHC provides.

The idea-bearing physicists might chat with some of their friends and colleagues to judge whether the question is worth pursuing. If they are students, they may consult with an adviser, usually a professor at their home university; if they are professors, they may hand off the idea to a student to work on. An idea that seems promising is then brought to one of the "working groups" each experiment has. The different working groups are devoted to various areas of interest: "top quarks" or "Higgs" or "exotics." (Exotics would include particles predicted by some of the speculative theories out there, or not predicted by anybody at all.) The working groups mull over the idea, after which the "convener" who leads the group makes a decision about whether it's worth moving forward with the analysis of this particular question. The experimentalists keep detailed Web pages that list each ongoing analysis, to help prevent duplication of effort—that's the reason the World Wide Web was invented.

Assuming an idea is given the nod by the relevant working group, the analysis moves forward. The physicist's life now alternates between working at a computer and participating in meetings, usually via videoconference. Doing an analysis is not by any means the only duty of an experimentalist; there is also hardware work, taking "shifts" overseeing the experiment as it's running, teaching (or taking) classes, giving talks, applying for grant money, and of course serving on committees and the thousand other bits of academic nonsense that are an inescapable part of university life. Occasionally the experimentalists are allowed to visit with their families or see the sun, but such frivolities are kept to a minimum.

At this point the data have been collected and safely stored on disk drives around the world; the job of an analyst is to turn that data into a useful physics result. It's rarely a matter of turning a crank. There are "cuts" to be made, throwing away some data that is noisy or irrelevant to the question being asked. (Maybe you want to look at events that feature two jets, but only with total energies greater than 40 GeV, and with an angle between them of at least 30 degrees.) Very often it is necessary to write specialized software to help tackle the specific problem under consideration. Data isn't very useful unless it can be compared with some theoretical expectation, so other pieces of software are used to calculate the predictions for what the data should look like according to different models. Even after cuts are applied to the data, it remains necessary to estimate the background noise that threatens to drown out your precious signal, which involves a give-and-take between calculations and other measurements. Throughout the process, regular updates are provided to the working group in charge, both in the form of written documentation and videoconference presentations.

Eventually one obtains a result. The next task is to convince the rest of the collaboration that your result is right—and nothing pleases a mob of cranky physicists like showing that someone else's analysis is wrong. Every project must first go through "preapproval" by the working group before eventually being approved by the collaboration as a whole. There is a committee whose sole job is to check that you've done your statistics correctly. The eventual goal is to publish a paper in a refereed journal, but the written paper must first be circulated throughout the collaboration, before ultimately being "blessed" by the publications committee. Only then can it be sent to a journal.

Nonscientists would be forgiven if they assumed that the author of a paper had actually written the paper. Of course, the person who writes the paper is *an* author, but everyone who contributes in an important way to the work being described is included on the list of authors. In experimental particle physics, the tradition is that every member of the collaboration is recognized as contributing to every paper produced by the experiment. You read that correctly: Every paper that comes out of CMS or ATLAS has more than three thousand authors. What's more, the authors are listed in alphabetical order, so that to an outsider it's completely impossible to determine who did the analysis or actually wrote the words in the paper. It's not an uncontroversial system, but it helps bring the collaboration together to stand behind every result they publish.

Generally, only after a paper is ready are the result of the analysis made public and the physicists involved permitted to give talks on the subject. The search for the Higgs boson is a special case, of course; everyone has known for years that this was a major goal for both experiments, and much of the preliminary groundwork was laid well ahead of time, allowing for the most rapid possible route from data to announcement. Still, until the experiments have verified that the data have been analyzed correctly, every effort is made to keep those results quiet.

I asked one physicist whether the results that ATLAS was getting were generally known within CMS, and vice versa. "Are you kidding?" I was told with a laugh. "Half of ATLAS is sleeping with half of CMS. Of course they know!" Superhuman levels of dedication to their craft notwithstanding, physicists are people too.

There are errors, and there are errors

The December updates on the Higgs search by Fabiola Gianotti and Guido Tonelli weren't the only seminars at CERN to garner public attention in 2011. In September of that year, Italian physicist Dario Autiero announced a result that ended up being more infamous than famous: neutrinos that appeared to be moving faster than the speed of light. The finding came from the OPERA experiment, which tracked neutrinos that were produced at CERN and traveled 450 miles underground to a detector in Italy. Because neutrinos interact so weakly, they can pass through many miles of solid rock with very little loss of intensity, making this kind of arrangement a uniquely effective window onto their properties.

The problem was obvious: Nothing is supposed to travel faster than light. Einstein figured that out, and it's one of the bedrock principles of modern physics. There are many good arguments in favor of this principle that had previously been verified in countless precision experiments. If it were to be overturned, it would be the most important finding in physics since the advent of quantum mechanics. We wouldn't have to start over completely from scratch, but new laws of nature would clearly be required. One worrisome consequence was that if you can go faster than light, you might also be able to travel backward in time, which instantly inspired a new genre of jokes. "The bartender says, 'We don't serve leptons here.' A neutrino walks into a bar."

Most physicists were immediately skeptical. On *Cosmic Variance* I wrote: "The things you need to know about this result are 1. It's enormously interesting if it's right. 2. It's probably not right." Even the OPERA collaboration members themselves seemed dubious of the implications of their findings, asking the physics community to help them understand why it might be incorrect. Of course, even the most confidently held theoretical belief must give way to an unimpeachable experimental result. The question was, how reliable was the result?

The OPERA finding was extremely statistically significant. The discrepancy between theory and observation was bigger than six sigma, more than strong enough to declare a discovery. Yet there were skeptics. And those skeptics were right. In March 2012, a different experiment, called ICARUS, attempted to replicate the OPERA findings but ended up with a very different result: that the neutrinos were completely consistent with the light-speed barrier.

Was this one of those cases where we just got preposterously (un)lucky, with a bizarre series of unlikely events conspiring to lead us astray? Not at all. The OPERA collaboration eventually pinpointed an important source of error in their original analysis, namely a loose cable that connected their master clock to a GPS receiver. The faulty cable led to a delay in the timing as measured by their detector, more than enough to account for the original anomaly. Once that was fixed, the effect went away.

The crucial lesson here is that sigmas aren't always enough. Statistics can help you decide how likely it is that your data are consistent with the null hypothesis, but only if your data are reliable in the first place. Scientists speak of "statistical errors" (because you don't have that much data, or there is intrinsic but random uncertainty in your measurements) and also "systematic errors" (due to some unknown effect that shifts the data uniformly in some direction). Just because you get a result that is statistically significant doesn't mean that it's true. This is a lesson taken very seriously by the physicists searching for the Higgs boson at the LHC.

Another issue is more murky: Were the OPERA physicists right to release their results to the world, and even to call a press conference at CERN about them? Arguments on either side have flown back and forth since the first announcement was made. On the one hand, the leaders of OPERA knew perfectly well that what they were claiming was astonishing, and they took the position that it was better to spread the news widely so that other scientists could help figure out whether something could have gone wrong. On the other, many people felt that the public image of science was hurt by the incident, first by raising the possibility that Einstein could have been wrong, and then admitting it was just a mistake. It could be a moot point; in an interconnected world where news travels rapidly, it may no longer be possible for large collaborations to keep surprising findings secret for very long.

Web 2.0

Tommaso Dorigo, a physicist on the CMS experiment and blogger at *A Quantum Diaries Survivor*, made a bold prediction in a 2009 talk to the World Conference of Science Journalists: The first time the outside world would hear about the final discovery of the Higgs boson, it would be through an anonymous comment left on a blog. In the end he wasn't exactly right, but close.

Prior to the Higgs boson, the last elementary particle in the Standard Model to be discovered was the top quark, pinned down by the Tevatron at Fermilab in 1995. That was about the same time that blogs were first coming into existence; the word "weblog" wasn't coined until 1997. There was no such thing as Facebook or Twitter; even MySpace, now long since condemned as hopelessly outdated, didn't start until 2003. The physicists working at the Tevatron might share some juicy gossip with other physicists, but there was not a lot of danger that a big discovery would go public ahead of time.

Things have changed. With the ease of communication on the Internet, anyone can spread news widely, and the ATLAS and CMS collaborations each have more than three thousand members. No matter how the leaders try to keep things under control, the chance that absolutely every one of them keeps knowledge of a major result to themselves is very slim indeed.

I will confess to being an enthusiastic proponent of blogs, although I try not to spread rumors that people don't want spread. I started blogging back in 2004 at a personal site called *Preposterous Universe*, and in 2005, switched to the group blog *Cosmic Variance*, which is now hosted by *Discover* magazine. The great thing about blogs is that they can be used for whatever purpose the author chooses. A wide variety of authors take full advantage of this freedom; just within the tiny subculture of blogs run by scientists and science writers, the examples range from the chatty and informal to the rigorous and mathematical, with everything from hard news to satire and gossip in between. Our goal at *Cosmic Variance* is to share interesting ideas and discoveries in science with a wide variety of readers, while allowing ourselves to muse and pontificate over whatever stirs our fancy. Some of our most popular posts have focused on the LHC, including a group effort live-blogging the startup in 2008 and the Higgs seminars in 2012.

One of my co-bloggers is John Conway, who is a professor of physics at the University of California, Davis, and an experimental physicist working on CMS. (JoAnne Hewett is another.) Conway's very first post, entitled "Bump Hunting," offered an insightful view of what it is like to be a working particle physicist. Sometimes the data can surprise you, and it's not always easy to tell whether you've stumbled on a world-changing discovery or are merely the victim of a statistical fluctuation.

Conway related the story of searching for the Higgs boson in Fermilab data (the LHC wasn't online yet) using his personal favorite channels, ones where a tau lepton is produced. They were doing a blind analysis of data from the CDF experiment at the Tevatron and finally got to the point where they were ready to open the box and see what was there. And the answer was . . . something was there! A small but unmistakable bump in the rate of producing two taus, the kind of thing you might expect from a Higgs boson with a mass of about 160 GeV. Only 2.5 sigma, but worth looking into. Most small bumps go away, but every real discovery begins with a small bump, so any breathing person in this situation would naturally get very excited. "The hair literally rose up on the back of my neck," he recalled.

In a follow-up post, Conway talked about the subsequent analysis, and revealed what he only learned later: Their sister experiment at Fermilab, known as "D Zero," actually saw a deficit where CDF was seeing an excess of events. That made it much less likely they were discovering a new particle. Further data didn't support the possibility that there was a new particle lurking there. But this story was a fantastic example of the emotional roller-coaster ride that is an inevitable part of life as an experimental scientist.

Sadly, not everyone interpreted it as such. An unfortunate number of readers got the impression that Fermilab had actually discovered the Higgs boson or something like it, and Conway had decided to spread the news by posting on our humble blog rather than writing a scientific paper and perhaps holding a press conference. This misimpression wasn't limited to overly enthusiastic commenters at our site; several journalists picked up on the prospect, which led to stories in *The Economist* and *New Scientist* and elsewhere. It was another cautionary lesson for physicists. People are extremely eager to hear anything and everything about the quest for the Higgs boson; great care is required to make sure that excitement is properly conveyed without giving people the impression that we've discovered more than we have.

Physics paparazzi

Giant particle accelerators aren't the only way to look for new physics. Payload for Antimatter Matter Exploration and Light-nuclei Astrophysics, or PAMELA for short, is an Italian cosmic-ray experiment that lives in low earth orbit, piggybacking on a Russian (nonmilitary) reconnaissance satellite. One of its main goals is to look for antimatter in cosmic rays, primarily positrons and antiprotons. It's not surprising that we see some antimatter; there are high-energy processes in outer space that occasionally produce antiparticles, just as there are at the LHC. What was surprising is that PAMELA observed substantially more positrons than they expected. This could be evidence for some astrophysical process that we don't currently understand, such as novel phenomena in neutron-star atmospheres; or it could be evidence for physics beyond the Standard Model, such as dark-matter particles annihilating and creating an excess of positrons. Various options are being investigated, although as time goes on the astrophysical options seem to be more promising.

Even more surprising, perhaps, is how the word of PAMELA's intriguing result got out. It often happens that a collaboration has preliminary results, not quite ready to be published or distributed, but enough to show to colleagues at a conference. That was the case with PAMELA at the International Conference on High Energy Physics in Philadelphia, in September 2008. PAMELA physicist Mirko Boezio briefly flashed a plot that showed an excess of positrons—a result that hadn't yet been incorporated into any publications.

Not briefly enough. While the plot was displayed, a young theorist named Marco Cirelli, sitting in the audience, quickly snapped a photo of it with his digital camera. Back home, he and collaborator Alessandro Strumia wrote a paper that proposed a new model of dark matter that could explain the excess, and sent it to the physics archive server http://arxiv.org, where it was distributed around the world. In their paper, they created plots in which they compare the theoretical prediction from their model with the data they extracted from their photo of the conference talk, with a footnote: "In order to comply with the publication policy, the preliminary data points for positron and antiproton fluxes plotted in our figures have been extracted from a photo of the slides taken during the talk."

Welcome to the new world. There is clearly something of a gray area here. A member of the collaboration might say that data that is not yet ready for publication should never be used in a theoretical analysis. But a member of the audience might reply that data that isn't ready shouldn't be shown in public talks, either. Piergiorgio Picozza, an Italian physicist and leader of PAMELA, was "very, very upset" that their data was acquired and used in this way. But Cirelli insists that he obtained permission from the PAMELA physicists who were at the conference: "We asked the PAMELA people [there], and they said it was not a problem."

As many teenagers have learned in the age of Facebook, anything you share with some people in the modern world you might as well share with everyone. Technology has made it effortless to distribute information, no matter how official or reliable that information may be. As Joe Lykken said in reference to yet another rumor, "Pre-blog, this sort of rumor would have circulated among perhaps a few dozen physicists. Now with blogs even string theorists who can't spell Higgs became immediately aware of inside information about [these] data."

Whispers

Rumors aren't always benign. In April 2011, an anonymous commenter on Peter Woit's blog *Not Even Wrong* leaked an internal ATLAS memo from Sau Lan Wu's team at Wisconsin. The contents were explosive, if true: strong evidence for a Higgs-like boson decaying into two photons. But it was too good to be true; to get that strong a signal with the relatively tiny amount of data they had at the time, the rate of Higgs decays would have had to be thirty times larger than the Standard Model prediction. Not impossible, but not something anyone was expecting, either. Unsurprisingly, once ATLAS did come out with an approved measurement, the signal had gone away.

This incident shows a downside of blogs. Internal memos such as this are the lifeblood of a large collaboration; they are written all the time, as part of the process described above for how an analysis matures into an approved result. Even the people who write the memos don't necessarily believe the result is real; they are simply pointing something out that deserves closer scrutiny. That's fine, as long as it is kept within the collaboration. If it goes public well before it's been vetted, there is a serious danger of misunderstanding, which can ultimately serve to undermine the public's confidence in the results that we do stand behind. Wu herself was furious: "Such a leak was totally unethical and irresponsible by the person who did it . . . The leak has damaged the freedom of conveying internal studies in written form to the collaborators. To me, this is an extremely sad affair."

In June 2012, CMS and ATLAS began to look carefully at the data they were able to collect so far that year. Everyone knew from the December 2011 seminars that there was a hint of Higgs at 125 GeV, so curiosity was understandably at a peak. As soon as the analysis began, rumors began to fly. There was a long-standing plan to present updates on the Higgs search in July at the ICHEP, scheduled for Melbourne. Chatter heated up when CERN announced that they weren't going to wait for Melbourne, but would instead have special seminars immediately beforehand in Geneva. Why would they do that if they weren't going to announce something big?

Things got so bad that Fabiola Gianotti, in an email to reporter Dennis Overbye at the *New York Times*, pleaded, "Please do not believe the blogs." But bloggers come in all stripes, and some tried to stem the tide rather than add to it. Michael Schmitt, a physicist at Northwestern and a member of CMS, wrote at his own at *Collider Blog*:

> My loyalty remains with my collaboration, especially the people who are working right now to carry out the analysis and verify the results, as well as to the people at the top who have to chart strategy and make difficult decisions. A little splash in a blog is not worth the bother it would cause all these people.

The undeniable truth is that, with six thousand people on the inside, someone is going to give in to the temptation to spill the beans—even before the beans have actually been collected and counted. One of the most frequent anti-blog complaints was not that results were being distributed ahead of time, but that the results didn't even yet exist; analysis takes time and often proceeds feverishly right up to the moment that a talk is given or a paper submitted for publication.

Meanwhile, others took the excitement and turned it into an opportunity to have a little fun. On June 20, various users of Twitter started passing back and forth satirical tweets about the Higgs. The hashtag #HiggsRumors even briefly became a "trending topic" on Twitter, an honor usually reserved for news involving *Jersey Shore* or Lady

Gaga. Jennifer Ouellette, a science writer and blogger (who is also my wife), collected some of the best tweets in a blog post.

@drskyskull: I hear the Higgs boson once shot a man just to watch him die. #HiggsRumors

@StephenSerjeant: ATLAS and CMS both beaten to Higgs detection by Chuck Norris. #HiggsRumors

@treelobsters: On the Summer Solstice, you can balance a Higgs Boson on end. #HiggsRumors

@tomroud: The God Particle actually is an atheist. #HiggsRumors

The best I could muster at the time was "Little Mikey from the LIFE cereal commercials died after eating Higgs bosons and drinking soda at the same time. #HiggsRumors." Probably this reveals more about my comedy skills (and my age) than anything else.

Hollywood squares

Los Angeles is an industry town, and the industry is entertainment. In early 2007, not long after I had first moved here, I got an unusual phone call. It was from Imagine Entertainment, the production company run by Ron Howard and Brian Grazer (*Apollo 13*, *A Beautiful Mind*, *The Da Vinci Code*). The filmmakers were in the planning stages for *Angels & Demons*, based on the Dan Brown book, which featured important scenes set at CERN. They wondered whether I'd be willing to drop by their Beverly Hills offices to chat about particle physics?

I admitted that I could probably fit it into my schedule. This was my first introduction to a little-known fact: Hollywood loves science.

It's the opposite of the usual stereotype, which is that movies and TV shows regularly serve up atrocious scientific mistakes, and typically portray scientists as either antisocial dweebs or mad geniuses bent on ruling the world. There certainly is a lot of that, but among many writers and directors there is a genuine interest in using honest science to improve the stories they want to tell. Howard and Grazer were sincerely interested in cosmology, antimatter, and the Higgs boson, and we shared an enjoyable lunch brainstorming ways to work physics into the film. Later, my wife, Jennifer, would become the first director of the Science and Entertainment Exchange, an effort from the National Academy of Sciences that works to improve interactions between scientists and Hollywood. Through the Exchange I was able to meet filmmakers like Ridley Scott, Michael Mann, and Kenneth Branagh, each of whom wanted to hear more about extra dimensions, time travel, and the Big Bang. Big-budget Hollywood movies are not meant to be documentaries or public service announcements for science; the storytelling always comes first, and suggestions from the scientists don't always make it into the final product. But many respected professionals who spin fairy tales on the silver screen appreciate the underlying wonder of scientific discovery.

For its own part, science isn't averse to going Hollywood to help its own cause. Science writer Kate McAlpine, who spent time at CERN working with ATLAS, in 2008, released a YouTube video entitled "Large Hadron Rap." The performance featured physicists dancing in front of LHC experiments while McAlpine rapped physics-themed lyrics over background beats:

Twenty-seven kilometers of tunnel underground
Designed with mind to send protons around
A circle that crosses through Switzerland and France
Sixty nations contribute to scientific advance
Two beams of protons swing round, through the ring they ride
'Til in the hearts of the detectors, they're made to collide
And all that energy packed in such a tiny bit of room
Becomes mass, particles created from the vacuum
And then . . .

Seven million views later, it is clear that the video struck a chord. There is no dearth of goofy YouTube videos on every subject under the sun; for some reason, this one stood out above the crowd. It serves as a reminder of how interested people can be in esoteric scientific questions when they are presented in a fun way.

The most ambitious project along these lines has been masterminded by David Kaplan, a particle theorist at Johns Hopkins University. Kaplan's day job is constructing models that can be tested at the LHC and in other experiments, but he has a long-standing interest in filmmaking. As a high school student, he remembers being academically unmotivated and didn't even apply to go to college. His sister, without telling him, sent an application in his name to Chapman University in California. To everyone's surprise he was accepted and spent a year there as a film major. It wasn't to his taste, and he ultimately ended up transferring to UC Berkeley and majoring in physics. He didn't go to graduate school immediately, partly because his Berkeley grades were so bad that he didn't think anyone would write him a letter of recommendation. Instead Kaplan moved to Seattle and earned money on the side by tutoring physics students at the University of Washington. After enough of the students compared him favorably to the graduate students at UW, he finally entered the PhD program there. All's well that ends well; he is now one of the leaders of a new generation of young particle theorists trying to push physics beyond the Standard Model.

As the LHC era crept up on us, Kaplan was struck by the unique nature of the moment in time. He would share with friends his impression that this was a make-or-break point in the history of science, if not the history of human intellectual development. If the LHC finds something interesting, it will launch us on a new path of discovery. If it doesn't, the prohibitive cost of modern particle physics might mean that this is the last major accelerator ever built. Kaplan became convinced that this high-stakes drama should be carefully documented. He would conduct interviews with particle physicists—both senior ones who have built careers on certain ideas about how nature works and would see them verified or thrown away, and younger ones who would have to cope with whatever the LHC did or did not reveal—and turn them into a book.

The problem was that even when it comes to scientific papers, Kaplan is a terribly laborious writer. The solution was obvious: Rather than write a book, he would make a movie. *Particle Fever* (the film's tentative title) was born.

As a new faculty member, Kaplan had been awarded a small fellowship from the Alfred P. Sloan Foundation. Usually such fellowships are used to fund computers or travel or some amount of support for graduate students. Instead, Kaplan got a TV director interested in his idea, and the two used the money to make a five-minute clip that could then be used to raise the serious money required to create a feature-length documentary film. Their original budget was $750,000 (since increased), and the real work began: raising money, hiring editors and writers, raising money, interviewing physicists, and raising money. They handed out small high-definition cameras to physicists at CERN who were able to record crucial events like the 2008 startup and the accident soon thereafter. Kaplan himself has devoted a substantial amount of time to the project. He gets no salary, and at one point his family had to give him a $50,000 loan to keep it afloat.

But interest has been immense. The development office at Johns Hopkins showed a clip to the university's board of directors, one of whom made an investment on the spot. The National Science Foundation, which supports much of the basic research in the United States and is constantly haranguing scientists to get more involved in public outreach, was thrilled to find out that one of their researchers was taking outreach seriously, and offered substantial support. Walter Murch, a highly respected Hollywood editor who has worked with George Lucas and Francis Ford Coppola and won multiple Academy Awards, became fascinated by the film and offered his services at well below his usual fee.

Throughout the process, Kaplan's goal has been to capture some of the quixotic fervor that pushes scientists to understand the universe just a tiny bit better than anyone has understood it before. The emotional stakes are high; physics is an experimental science, and the most brilliant theorists in the world get little credit if the theory they propose turns out not to be the path nature has chosen. In Kaplan's words,

> In the end, it's an incredibly heroic exercise. And it is filled with different egos, and intensity, and overconfidence maybe. But what you understand is that people fool themselves. Scientists create a world in their brain, in order to get themselves to work as hard as they do and to keep going, knowing that it could be a complete failure. Their entire career could just be in the toilet as totally irrelevant.

As of mid-2012, *Particle Fever* is nearing completion, and the team is hoping to get chosen for the Sundance Film Festival in January 2013. Fittingly, they are wildly ambitious, hoping for an eventual wide theatrical release that will truly bring the LHC to the masses. Whether that succeeds, they will certainly have created a singular document that will stand as a testament to both the excitement and the nervousness of physicists at the dawn of the LHC era.

And David Kaplan will be able to devote himself to physics full-time once again. As interesting and novel as the process was, there's no danger he will be changing jobs anytime soon:

> Making a movie is just a terrible experience. It's so illogical, and there's ego, and people making arguments in ways that just don't make any sense. I hate it . . . I love physics.

ELEVEN
NOBEL DREAMS

In which we relate the fascinating tale of how the "Higgs" mechanism was invented and think about how history will remember it.

It was 1940, and Germany had just invaded Denmark. Niels Bohr, one of the founders of quantum mechanics and director of the Institute for Theoretical Physics in Copenhagen, was in possession of valuable pieces of contraband he needed to keep hidden from the Nazis at all costs: two gold medals that accompanied winning the Nobel Prize. How could he keep them away from the approaching army?

Bohr had won the Nobel in 1922, but neither of the medals belonged to him; he had previously auctioned off his prize medal to help support resistance forces in Finland. They belonged to Max von Laue and James Franck, two German physicists, who had illegally smuggled their medals (which were engraved with their names) out of the country to keep them away from the Nazis. Bohr turned to his friend, the chemist George de Hevesy, who hit upon a brilliant idea: They would dissolve the medals in acid. Gold doesn't dissolve easily, so the scientists turned to aqua regia, a highly corrosive mixture of nitric acid and hydrochloric acid, renowned for its ability to tear down "noble" metals. Placed in the aqua regia, over the course of an afternoon, the Nobel medals gradually dissociated into their individual atoms, which remained suspended in the solution. Any soldiers that would come poking around looking for suspicious hidden treasure would find nothing but a couple of innocuous flasks of chemicals hidden among hundreds of similar-looking containers.

The ruse worked. After the war, scientists were able to recover the gold by precipitating the atoms out of de Hevesy's solution. Bohr delivered the metal back to the Royal Swedish Academy of Sciences in Stockholm, which was able to recast von Laue's and Franck's Nobel medals. De Hevesy himself, who fled to Sweden in 1943, won the Nobel Prize in Chemistry in 1944—not for discovering new techniques in hiding contraband, but for the use of isotopes in tracing chemical reactions.

In case it wasn't obvious, people take Nobel Prizes very seriously. At the end of the nineteenth century, chemist Alfred Nobel, the inventor of dynamite, established prizes in Physics, Chemistry, Physiology or Medicine, Literature, and Peace, which have been awarded each year since 1901. (The Economics prize, begun in 1968, is run by a different organization.) Nobel passed away in 1896, and the executors of his will were surprised to find that he had donated 94 percent of his considerable fortune to the establishment of the prizes.

In the years since, the Nobel Prizes have become universally recognized as the pinnacle of scientific recognition. That isn't quite the same as scientific "achievement"—the Nobels have quite specific criteria, and there are endless arguments about how well the prizes match up with the truly important scientific discoveries. Nobel's original will aimed the prizes at "those who, during the preceding year, shall have conferred the greatest benefit on mankind," and the Physics prize in particular "to the person who shall have made the most important 'discovery' or 'invention' within the field of physics." To some extent these instructions are simply ignored; after a few early prizes were given to findings that later turned out to be in error, nobody pretends anymore that the prizes recognize work done in the preceding year. Crucially, making a "discovery" is not the same as being recognized as one of the world's leading scientists. Some discoveries are made somewhat by accident, by people who later leave the field. And some scientists do fantastic work over the course of a lifetime, but don't quite have a single world-changing discovery that rises to the level of a Nobel.

There are other criteria that highly constrain the Nobel choices. Prizes are not awarded posthumously, although if a laureate passes away between the time when the decision is made and when it is announced, the prize is still given to them. Most important for physics, the prize is not given to more than three people in any one year. Unlike the Peace prize, for example, the Physics prize isn't given to an organization or a collaboration; it is given to three or fewer individuals. That poses something of a challenge in the Big Science era.

When it comes to theoretical contributions, it's not enough to be smart, or even to be right; you have to be right, and your theory has to be confirmed by experiment. Stephen Hawking's most important contribution to science is the realization that black holes give off radiation due to the effects of quantum mechanics. The large majority of physicists believe he is right, but at this point it's a purely theoretical result; we haven't observed any evaporating black holes, and we don't have any promising way of doing so with current technologies. It's quite possible that Hawking will never win the Nobel Prize, despite his incredibly impressive contributions.

To outsiders, it can sometimes seem like the whole point of doing research is to win the Nobel Prize. That's not the case; the Nobel captures important moments in science, but scientists themselves recognize there is a rich tapestry of progress that includes many contributions, great and small, which build on one another over the years. Still—let's admit it—winning the Nobel is a big deal, and physicists certainly keep track of which discoveries might someday qualify.

There is no question that discovering the Higgs boson is the kind of achievement that is certainly worthy of the Nobel Prize. For that matter, inventing the theory that predicted the Higgs in the first place is undoubtedly prize-worthy. But that doesn't necessarily imply that any prizes are actually going to be given. Who might win them? Ultimately it's not prizes that matter, it's the science; but we have a good excuse for looking at the fascinating history of the ideas behind the Higgs boson and how physicists set about searching for it. The goal of this chapter is not to provide a definitive history nor to adjudicate who deserves what prize. Quite the opposite: By looking at how the ideas developed over time, it should become clear that the Higgs mechanism, like many great ideas in science, involved many crucial steps to the final answer. Attempting to draw a bright line between three (or fewer) people who deserve a prize and the many others who don't necessarily does great violence to the reality of the development, even if it does make for good news copy.

In this chapter we're going to try to get the history right, although such a brief account will necessarily be incomplete. For history, however, the details often matter. Therefore, compared with the rest of the book, this chapter will go a little bit more into technical details. Feel free to skip over it, if you don't mind missing out on some fascinating physics and compelling human drama.

Superconductivity

In Chapter Eight we explored the deep connection between symmetries and forces of nature. If we have a "local" or "gauge" symmetry—one that operates independently at each point in space—it necessarily comes with a connection field, and connection fields give rise to forces. This is how gravity and electromagnetism both work, and in the 1950s, Yang and Mills suggested a way to extend the idea to other forces of nature. The problem, as Wolfgang Pauli forcibly pointed out, is that the underlying symmetry always comes associated with massless boson particles. That's part of the power of symmetries: They imply stringent restrictions on the properties that particles can have. The symmetry underlying electromagnetism, for example, implies that electric charge is exactly conserved.

But forces mediated by massless particles—as far as anyone knew at the time—stretch over infinite distances and should be very easy to detect. Gravity and electromagnetism are the obvious examples, while the nuclear forces seem very different. Today we recognize that the strong and weak interactions are also Yang-Mills-type forces, with the massless particles hidden from us for different reasons: In the strong force the gluons are massless but confined inside hadrons, while in the weak force the W and Z bosons become massive because of spontaneous symmetry breaking.

Back in 1949, American physicist Julian Schwinger had put forward an argument that forces based on symmetries would always be carried by massless particles. He kept thinking about the problem, however, and in 1961, he realized that his argument was not airtight: There was a loophole that allowed for the gauge bosons to get a mass. He wasn't quite sure how it might actually happen, but he wrote a paper that pointed out his previous mistake. Schwinger was famously elegant and precise in his personal style as well as his physics research. He stood in contrast with Richard Feynman, with whom he and Sin-Itiro Tomonaga shared the Nobel Prize in 1965. Feynman was known for his boisterously informal personality and deeply intuitive approach to physics, while Schwinger was unfailingly meticulous and proper. When he wrote a paper pointing out a flaw in a well-accepted piece of conventional wisdom, people took him seriously.

The question remained: What could cause the force-carrying bosons to get a mass? The answer came from a slightly unexpected source: not particle physics but condensed matter physics, the study of materials and their properties. In particular, insights borrowed from the theory of superconductors—materials with no resistance to electricity, such as those that power the giant magnets in the LHC.

Electrical current is the flow of electrons through a medium. In an ordinary conductor, the electrons keep bumping into atoms and other electrons, providing resistance to the flow. Superconductors are materials in which, when the temperature is low enough, current can flow through unimpeded. The first good theory of superconductors was put forward by Soviet physicists Vitaly Ginzburg and Lev Landau in 1950. They suggested that a special kind of field permeates the superconductor, which acts to give a mass to the ordinarily massless photon. They weren't necessarily thinking of a new fundamental field of nature, but a collective motion of electrons, atoms, and electromagnetic fields—much like a sound wave doesn't come from vibrations of a fundamental field, but from the collective motion of atoms in the air bumping into one another.

Although Landau and Ginzburg proposed that some kind of field was responsible for superconductivity, they didn't specify what that field actually was. That step was carried out by American physicists John Bardeen, Leon Cooper, and Robert Schrieffer, who invented what's called the "BCS theory" of superconductivity in 1957. The BCS theory is one of the milestones of twentieth-century physics, and certainly deserves a book of its own. (This isn't that book.)

BCS borrowed an idea of Cooper's, that pairs of particles could team up at very low temperatures. It's these "Cooper pairs" that make up the mysterious field suggested by Landau and Ginzburg. While a single electron would continually meet resistance by bumping into the atoms around it, a Cooper pair can combine in a clever way so that every nudge that pushes on one electron exerts an equal and opposite pull on the other one (and vice versa). As a result, the paired electrons glide through the superconductor unimpeded.

This is directly related to the fact that photons are effectively massive inside the superconductor. When particles are massless, their energy is directly proportional to their velocity and can range from zero up to any number you imagine. Massive particles, by contrast, come with the minimum energy they can possibly have: their rest energy, given by $E = mc^2$. When moving electrons are jostled by atoms and other electrons in a material, their electric field gently shakes, which creates very low-energy photons you would hardly ever notice. It's that continual emission of photons that lets the electrons lose energy and slow down, diluting the

Joe Incandela, spokesperson for CMS in 2012.

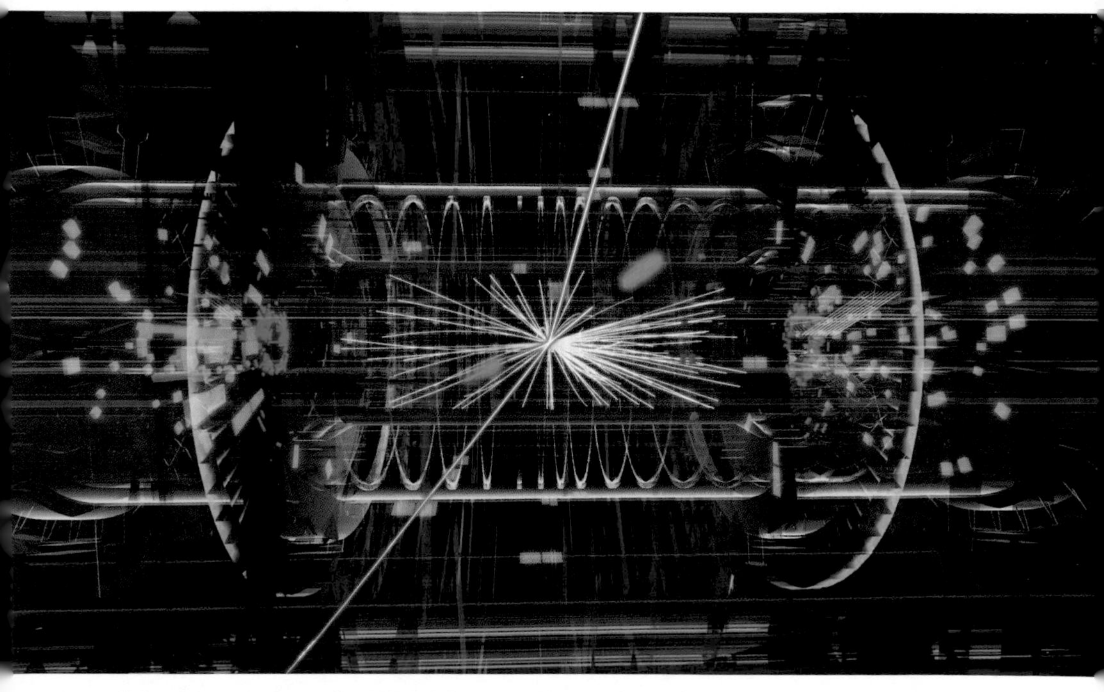

A candidate Higgs event at the ATLAS detector. The two long blue lines are muons, and the short blue lines are electrons, so this could represent the decay of the Higgs into two Z bosons.

The ATLAS detector under construction. Note the person standing at the middle bottom. The eight giant tubes are magnets used to deflect muons in order to measure their energies.

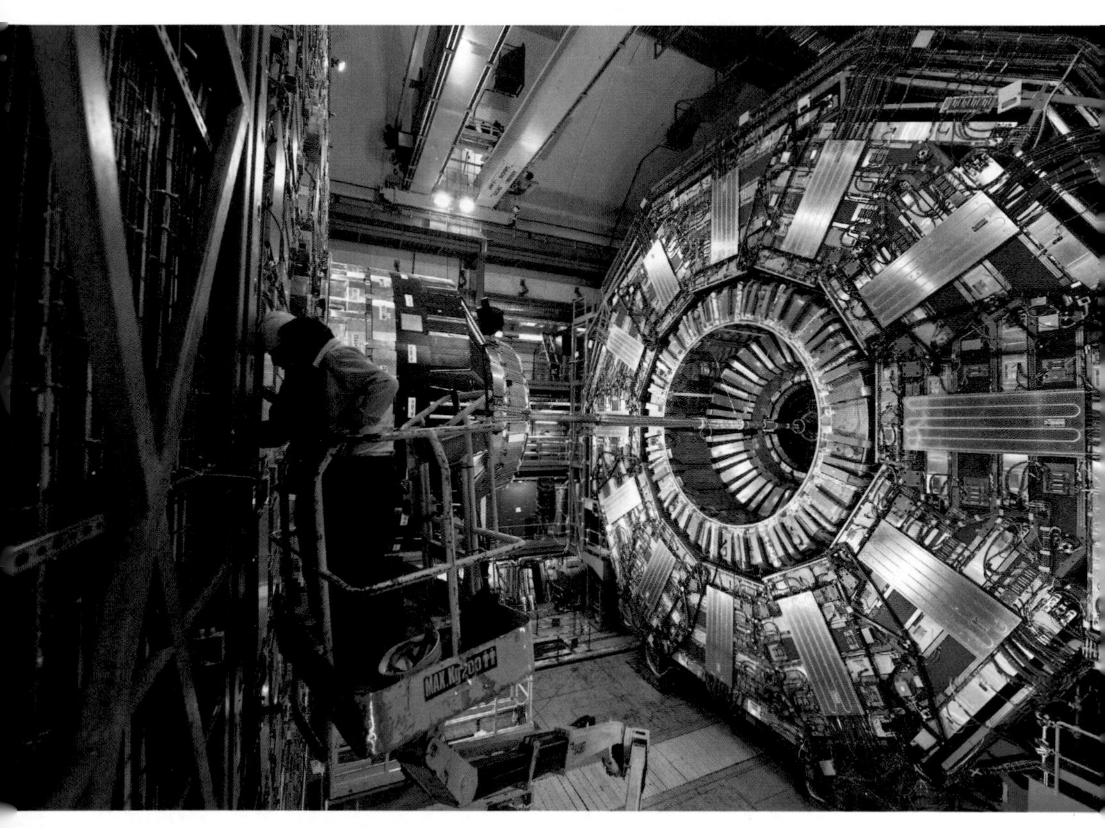

The CMS detector under construction.

Yoichiro Nambu, pioneer of symmetry breaking, gluons, and string theory.

CREATIVE COMMONS/BESTYTHEDEVINE

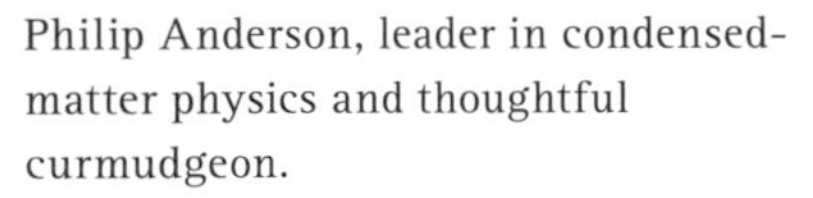

Philip Anderson, leader in condensed-matter physics and thoughtful curmudgeon.

CREATIVE COMMONS/PHILIP WARREN

Left to right: Tom Kibble, Gerald Guralnik, Carl Richard Hagen, François Englert, and Robert Brout, at the 2010 Sakurai Prize ceremony. Peter Higgs shared the award but was absent.

Peter Higgs, visiting the ATLAS experiment.

Left to right: Sheldon Glashow, Abdus Salam, and Steven Weinberg, at the 1979 Nobel Prize ceremony.

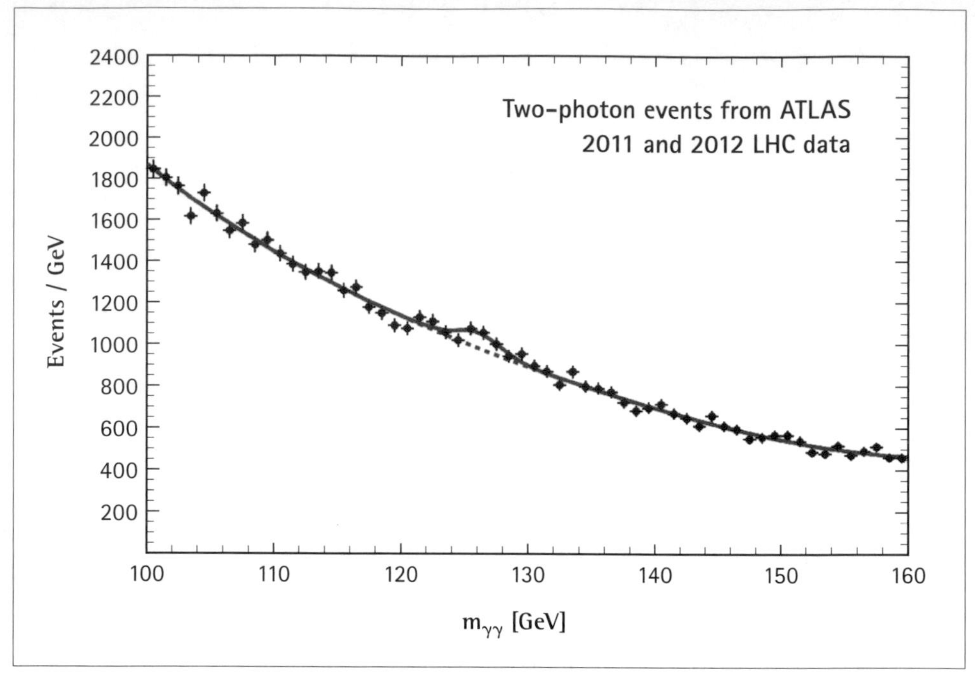

The data produced and analyzed in the LHC's search for the Higgs. These plots show the number of events that produce two high-energy photons, where the total energy of the photons ranges from 100 to 160 GeV, in the 2011–2012 data from ATLAS and CMS. The dotted lines shows the prediction without any Higgs boson; the solid curve includes a Higgs with a mass of 126.5 GeV (ATLAS) or 125.3 GeV (CMS).

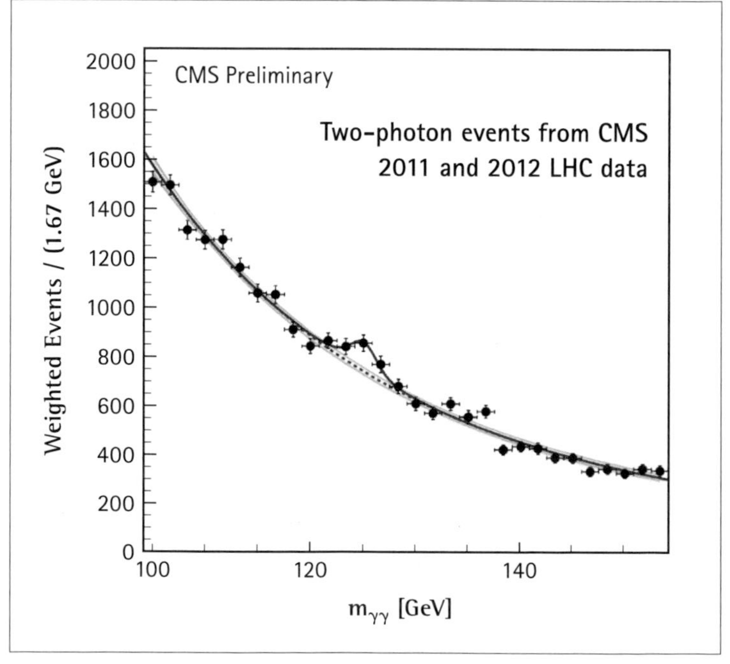

Why we do science.

ZACH WEINERSMITH, SATURDAY MORNING BREAKFAST CEREAL

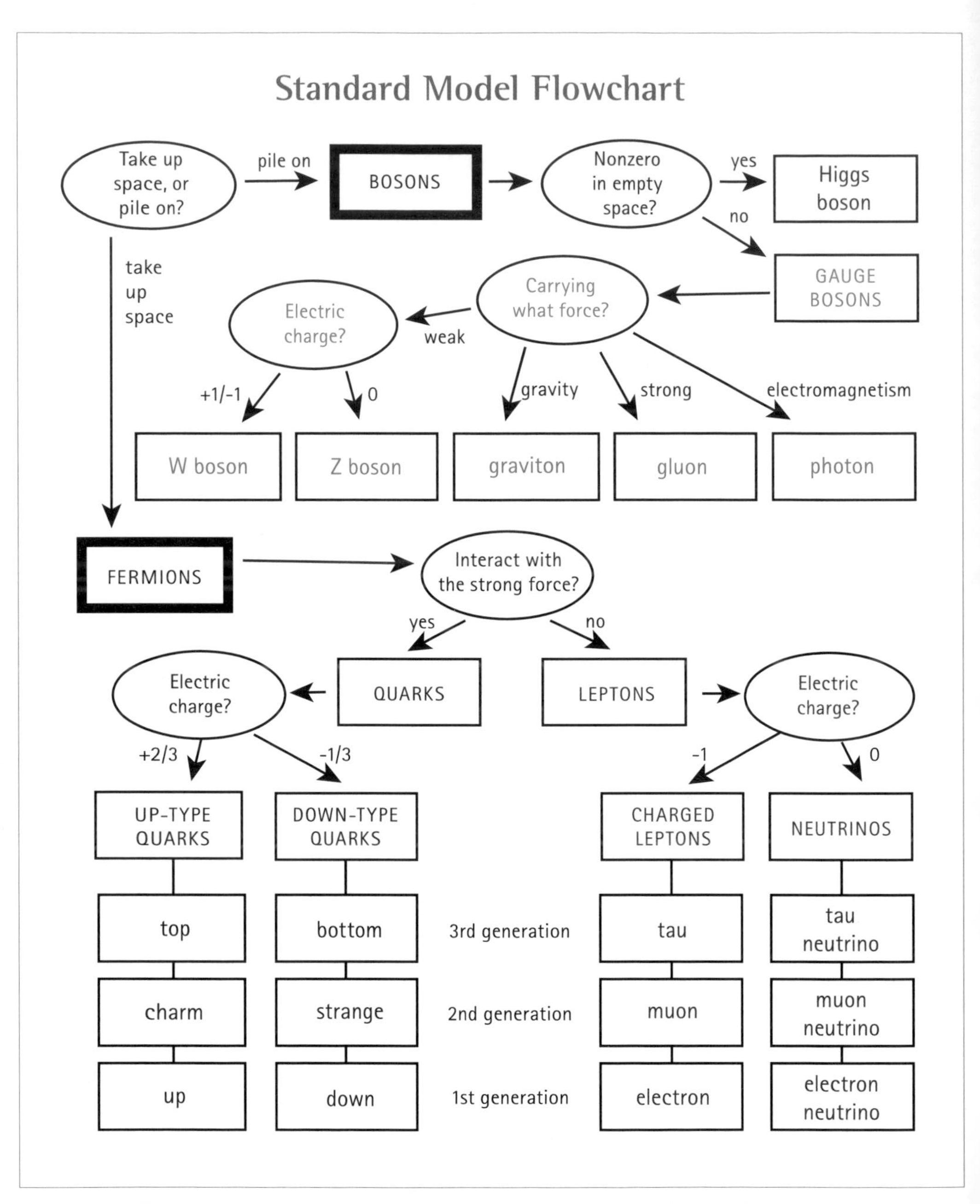

A flowchart illustrating the elementary particles of the Standard Model. This is the modern version of the periodic table of the elements. Quarks are in blue, leptons in purple, gauge bosons in green, and the Higgs boson in red.

SEAN CARROLL

current. Because photons obtain a mass in the Landau-Ginzburg and BCS theories, there is a certain minimum energy required to make them. Electrons that don't have enough energy can't make any photons, and therefore can't lose energy: The Cooper pairs flow through the material with zero resistance.

Electrons, of course, are fermions, not bosons. But when they come together to make Cooper pairs, the result forms a boson. We have defined bosons as force-carrying fields that can pile up, as opposed to fermions, which are matter fields that take up space. As we discuss in Appendix One, fields have a property called "spin" that also distinguishes bosons from fermions. All bosons have spins that are whole numbers: 0, 1, 2 . . . Fermions, meanwhile, have spins that are whole numbers plus one-half: 1/2, 3/2, 5/2 . . . The electron is a fermion with spin equal to 1/2. When particles get together, their spins can either add or subtract; so a pair of two electrons can have either spin-0 or -1—just right for making bosons.

This introduction is deeply unfair to the intricacies of the Landau-Ginzburg and BCS theories, which tell a rich story of many kinds of particles moving together in an intrinsically quantum-mechanical way. For our present purposes, the take-home message is straightforward: A bosonic field pervading space can give a mass to photons.

Spontaneous symmetry breaking

That last statement sounds pretty close to the Higgs idea. But a puzzle remained: How do we reconcile the idea that photons have mass inside a superconductor with the conviction that the underlying symmetry of electromagnetism forces the photon to be massless?

This problem was tackled by a number of people, including American physicist Philip Anderson, Soviet physicist Nikolay Bogolyubov, and Japanese-American physicist Yoichiro Nambu. The key turned out to be that the symmetry was indeed there, but that it was *hidden* by a field that took on a nonzero value in the superconductor. According to the jargon that accompanies this phenomenon, we say the symmetry is "spontaneously broken": The symmetry is there in the underlying equations, but the particular solution to those equations in which we are interested doesn't look very symmetrical.

Yoichiro Nambu, despite the fact that he won the Nobel Prize in 2008 and has garnered numerous other honors over the years, remains relatively unknown outside physics. That's a shame, as his contributions are comparable to those of better-known colleagues. Not only was he one of the first to understand spontaneous symmetry breaking in particle physics, he was also the first to propose that quarks carry color, to suggest the existence of gluons, and to point out that certain particle properties could be explained by imagining that the particles were really tiny strings, thus launching string theory. Theoretical physicists admire Nambu's accomplishments, but his inclination is to avoid the limelight.

Nambu's office was across the hall from mine while I was a faculty member at the University of Chicago. We didn't interact much, but when we did he was unfailingly gracious and polite. My major encounter with him was one time when he knocked on my door, hoping that I could help him with the email system on the theory group computers, which tended to take time off at unpredictable intervals. I wasn't much help, but he took it philosophically. Peter Freund, another theorist at Chicago, describes Nambu as a "magician": "He suddenly pulls a whole array of rabbits out of his hat, and before you know it, the rabbits reassemble in an entirely novel formation and by God, they balance the impossible on their fluffy cottontails." His highly developed sense of etiquette, however, failed him when he was briefly appointed as department chair: Reluctant to explicitly say no to any question, he would indicate disapproval by pausing before saying yes. This led to a certain amount of consternation among his colleagues, once they realized that their requests hadn't actually been granted.

After the BCS theory was proposed, Nambu began to study the

phenomenon from the perspective of a particle physicist. He put his finger on the key role played by spontaneous symmetry breaking and began to wonder about its wider applicability. One of Nambu's breakthroughs was to show (partly in collaboration with Italian physicist Giovanni Jona-Lasinio) how spontaneous symmetry breaking could happen even if you weren't inside a superconductor. It could happen in *empty space*, in the presence of a field with a nonzero value—a clear precursor to the Higgs field. Interestingly, this theory also showed how a fermion field could start out massless but gain mass through the process of symmetry breaking.

As brilliant as it was, Nambu's suggestion of spontaneous symmetry breaking came with a price. While his models gave masses to fermions, they also predicted a new massless boson particle—exactly what particle physicists were trying to avoid, since they didn't see any such particles created by the nuclear forces. These weren't gauge bosons, since Nambu was considering the spontaneous breakdown of global symmetries rather than local ones; these were a new kind of massless particle. Soon thereafter, Scottish physicist Jeffrey Goldstone argued that this wasn't just an annoyance: Spontaneously breaking a global symmetry always gives rise to massless particles, now called "Nambu-Goldstone bosons." Pakistani physicist Abdus Salam and American physicist Steven Weinberg then collaborated with Goldstone in promoting this argument to what seemed like an airtight proof, now called "Goldstone's theorem."

One question that must be addressed by any theory of broken symmetry is, what is the field that breaks the symmetry? In a superconductor the role is played by the Cooper pairs, composite states of electrons. In the Nambu–Jona-Lasinio model, a similar effect happens with composite nucleons. Starting with Goldstone's 1961 paper, however, physicists became comfortable with the idea of simply positing a set of new fundamental boson fields whose job it was to break symmetries by taking on a nonzero value in empty space. The kind of fields required are known as a "scalar" fields, which is a way of saying they have no

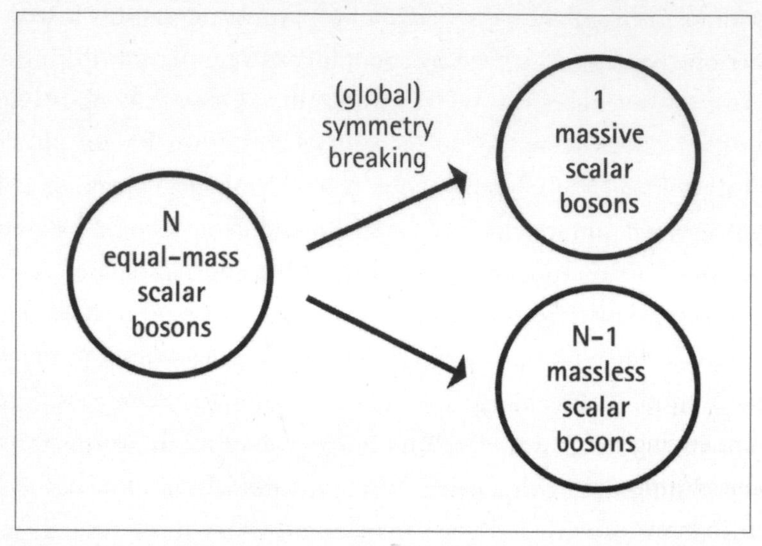

What happens when you spontaneously break a global symmetry. Without symmetry breaking, there would be a certain number N of scalar bosons with equal masses. After the symmetry is broken, all but one of them become massless Nambu-Goldstone bosons. The remaining one is massive.

intrinsic spin. The gauge fields that carry forces, although they are also bosons, have spin-1, except for the graviton, which is spin-2.

If the symmetry wasn't broken, all the fields in Goldstone's model would behave in exactly the same way, as massive scalar bosons, due to the requirements of the symmetry. When the symmetry is broken, the fields differentiate themselves. In the case of a global symmetry (a single transformation all throughout space), which is what Goldstone considered, one field remains massive, while the others become massless Nambu-Goldstone bosons—that's Goldstone's theorem.

Reconciliation

This was bad news. It seemed as if, even if you followed BCS and Nambu to use spontaneous symmetry breaking as a way to give mass to the hypothetical Yang-Mills bosons that could carry the nuclear forces, the very technique you employed gave rise to another kind of massless boson that wasn't seen in experiments.

Fortunately, the resolution to this puzzle was known almost as soon as the puzzle arose. At least it was known to Phil Anderson at Bell Labs, and he tried his best to share it with the world. Anderson, who won the Nobel in 1977, is recognized as one of the world's leading condensed-matter physicists. He has been a vocal champion for the intellectual status of condensed matter as a field; his celebrated 1972 article entitled "More Is Different" helped spread the word that studying the collective behavior of many particles was at least as interesting and fundamental as studying the underlying laws obeyed by the particles themselves. In contrast to the reticent Nambu, Anderson has always been willing to speak his mind, often in provocative ways. The subtitle of a collection of his essays is "Notes from a Thoughtful Curmudgeon," and the biography on the back flap informs us that "at press time he was involved in several scientific controversies about high profile subjects, in which his point of view, though unpopular at the moment, is likely to prevail eventually."

While Nambu was certainly inspired by the BCS theory, the model he and Jona-Lasinio proposed of spontaneous symmetry breaking in empty space featured a global symmetry, not a local (gauge) symmetry. It's local symmetries that give rise to connection fields, and therefore to forces of nature. Global symmetries help us to understand the presence or absence of different interactions, but they don't lead to new forces.

Anderson was not a particle physicist, but he understood the basic ideas behind Nambu-Goldstone bosons; they played an important (if somewhat implicit) role in his work on the BCS theory in 1958. He had discussed the dynamical consequences of symmetry breaking as early as 1952; today he considers this insight to be his biggest contribution to physics. Anderson also knew that it couldn't really be true that spontaneous symmetry breaking was always associated with massless particles, because spontaneous symmetry breaking occurred in the BCS model, and that model didn't have any massless particles.

So in 1962, prompted by Schwinger's admission from a year earlier,

Anderson wrote a paper (published in 1963) that attempted to explain to particle physicists how to avoid the menace of the massless particles. It was an elegant solution: The massless force-carrying particles you start with, and the massless Nambu-Goldstone bosons given to you by spontaneous symmetry breaking, *combine* to form a single massive force-carrying particle. This is otherwise known as "two wrongs make a right."

Anderson is explicit about the import of his analysis:

> It is likely, then, considering the superconducting analog, that the way is now open for a degenerate-vacuum theory of the Nambu type without any difficulties involving either zero-mass Yang-Mills gauge bosons or zero-mass Goldstone bosons. These two types of bosons seem capable of "canceling each other out" and leaving finite mass bosons only.

Despite this analysis, however, particle physicists did not get the message. Or they got the message but didn't believe it. Anderson's argument concerned the general properties of fields in the presence of spontaneous breakdown of a gauge symmetry, but he didn't write down an explicit model with a fundamental field that did the symmetry breaking. He showed that the conclusions of Goldstone's theorem were avoided, but he didn't explain precisely what had gone wrong with the assumptions of the theorem.

Most important, in condensed matter systems it's easy to measure your velocity with respect to the material you are in. In empty space, however, there is no preferred frame of rest; relativity assures us that all velocities are created equal. In the proofs of Goldstone's theorem, relativity played a crucial role. To many particle physicists, the fact that Goldstone had a rigorously proven theorem seemed to trump Anderson's examples to the contrary, and they appealed to relativity to reconcile the differences. In 1963, Harvard physicist Walter Gilbert wrote a paper that put forward this argument explicitly. (Gilbert was in the process of leaving particle physics for biology. The career switch wasn't necessitated by any lack of talent; in 1980, he shared the Nobel Prize in Chemistry for his work on nucleotides.) A 1964 paper by Abraham Klein and Benjamin Lee studied how the Goldstone theorem could be avoided in the nonrelativistic context, and suggested that similar reasoning would work equally well when relativity was included, but their arguments weren't considered definitive.

Anderson himself was leery of taking the notion of spontaneous symmetry breaking in empty space too seriously, for a good reason that nags at us to this day. If you have some field with a nonzero value in empty space, we expect that field will carry energy. It could be a positive amount of energy or a negative amount, but there's no special reason for it to be zero. Einstein taught us long ago that energy in empty space—vacuum energy—has an important effect on gravity, pushing or pulling on the expansion of the universe (depending on whether the energy is positive or negative). A simple back-of-the-envelope calculation reveals that the energy we're talking about is so enormously large that we would have noticed it long ago—or, more accurately, we wouldn't be around to notice it, as the universe would have blown apart or recollapsed shortly after the Big Bang. This is the "cosmological constant problem," which remains one of the most pressing questions in theoretical physics. These days we believe that there very likely *is* a tiny, positive energy in empty space, the "dark energy" that makes the universe accelerate, for which the Nobel Prize in Physics was awarded in 2011. But the numerical amount of that dark energy is much smaller than we had any right to expect, so the mystery remains.

1964: Englert and Brout

Every physicist, when in possession of that precious commodity called "a good idea," lives in fear of being scooped—of having their idea occur to someone else and get published before they get around to it. Given the number of ideas it is possible to have, you might expect that this is a rare event. But ideas don't appear at random; all scientists are embedded in a communication mosaic of talks and papers and informal conversation, and it's very common that two or more people who have never met each other will nevertheless be thinking about the same problems. (In the seventeenth century, Isaac Newton and Gottfried Leibniz both managed to invent calculus without coordinating ahead of time.)

In 1964, the year the Beatles took America by storm, three independent groups of physicists came up with very similar proposals that showed how spontaneous breakdown of a local symmetry doesn't produce any massless bosons at all, only massive ones that lead to short-range forces. The first to appear was a paper by François Englert and Robert Brout of the Université libre de Bruxelles in Belgium. Next up was Peter Higgs from Edinburgh, Scotland, with two papers in rapid succession. And then Americans Carl Richard Hagen and Gerald Guralnik (who had been Walter Gilbert's PhD student) teamed up with Englishman Tom Kibble to write a paper. All three groups worked independently, and all three deserve some of the credit for inventing what we now know as the "Higgs mechanism"—but the very precise apportionment of credit continues to be debated.

The Englert and Brout paper was short and to the point. The two physicists had met in 1959, when Englert came to Cornell as a postdoc to work with Brout. The first day they met, they went out for a drink, which turned into several drinks, as the two hit it off immediately. When Englert returned to Belgium in 1961 to take up a faculty position there, Brout and his wife arranged a temporary visit to Brussels and soon decided to stay there for good. They remained close friends and collaborators until Brout passed away in 2011.

They have two kinds of fields in their discussion: the force-carrying gauge boson and a set of two symmetry-breaking scalar fields that take on a nonzero value in empty space. It's a similar setup as in Goldstone's work on global symmetry breaking, with the addition of the gauge field required by a local symmetry. But they don't devote much attention to the properties of the scalar fields, concentrating instead on what happens to the gauge field. They show, using Feynman diagrams, that it gets a mass without violating the underlying symmetry—perfectly in accord with the requirements of relativity, and in contradiction to Gilbert's worry. All this was done apparently without knowing anything of Anderson's paper from the previous year.

1964: Higgs

Peter Higgs, after receiving a PhD from University College London, returned to his native Scotland to take up a lectureship at the University of Edinburgh in 1960. He was aware of Anderson's work and was interested in showing explicitly how Goldstone's theorem could be avoided in a relativistic theory. In June 1964, Higgs opened the latest issue of *Physical Review Letters* (*PRL*), the premier physics journal from the United States, and came across Gilbert's paper. He later recalled: "I think my reaction was to say 'shit' because he seemed to have closed the door on the Nambu programme." But Higgs didn't give up. He remembered that Schwinger had found a loophole in the usual argument that gauge bosons must be massless because of symmetry considerations, and thought it should be possible to extend the loophole to the case of spontaneously broken symmetries. Realizing that these were important issues, Higgs quickly wrote a short paper that was published in *Physics Letters*, the European counterpart to *Physical Review Letters*. Here, for the first time, it was shown explicitly how the assumptions behind Goldstone's theorem can be sidestepped in the case of a gauge symmetry, even when relativity is completely respected.

What Higgs didn't have in that first paper was a specific model in which the massless bosons were actually eradicated. In the second paper he provided exactly that, examining the behavior of a Goldstone-style pair of symmetry-breaking scalar fields coupled to a force-carrying gauge field, showing that the gauge field gobbled up the Nambu-Goldstone boson to make a single massive gauge boson. He sent this second paper to *Physics Letters* again—where it was promptly rejected. This was surprising to Higgs, who couldn't understand why a journal would publish a paper saying, "Massive gauge bosons are possible" but not one saying, "Here is an actual model with massive gauge bosons." But once again Higgs refused to give up; he added a couple of paragraphs elaborating on the physical consequences of the model, and sent it off to *Physical Review Letters* in the United States, where it was accepted. The reviewers there—who Higgs later learned was Nambu—suggested adding a reference to the Englert and Brout paper, which had just appeared.

Among the additions Higgs made after his second paper was rejected was a comment noting that his model didn't only make the gauge bosons massive, it also predicted the existence of a massive scalar boson—the first explicit appearance of what we now know and love as the "Higgs boson." Remember that Goldstone's model of broken global symmetry predicted a number of massless Nambu-Goldstone bosons, but also a leftover massive scalar. In the case of a local symmetry, the would-be massless scalar bosons get eaten by the gauge fields, which become massive. But the massive scalar field from Goldstone's theory is still there in Higgs's theory. Englert and Brout didn't discuss this other particle, although in hindsight it's implicit in their equations (as it was in Anderson's work).

Looking ahead a bit, in the real-world implementation of the Higgs mechanism in the Standard Model, before symmetry breaking we start with four scalar bosons and three massless gauge bosons. When the symmetry is broken by the scalars getting a nonzero value in empty space, three of the scalar bosons are eaten by the gauge bosons. We're left with three massive gauge bosons: the Ws and the Z, and one massive scalar—the Higgs. Another gauge boson starts off massless and stays that way—that's the photon. (The photon is actually a mixture of some of the original gauge bosons, but this is getting complicated enough.) In one sense, we discovered three-quarters of the Higgs bosons out in the 1980s, when we found the massive Ws and Z.

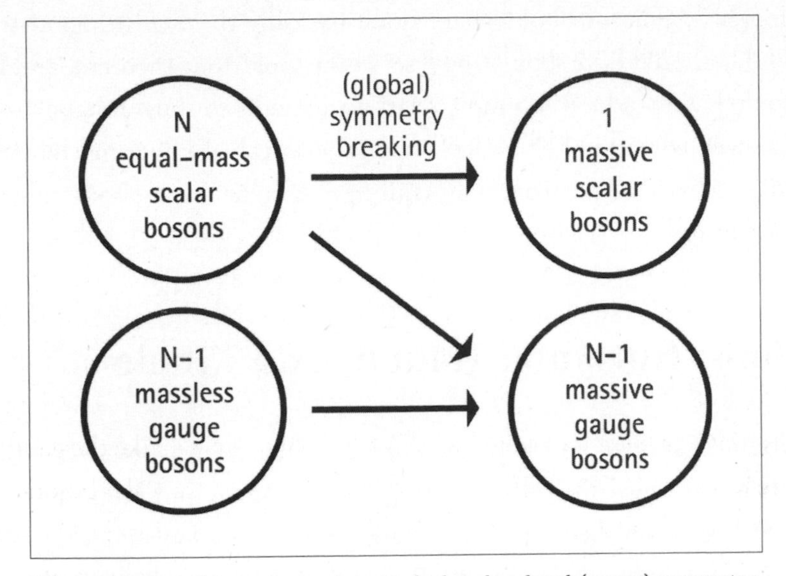

What happens when you spontaneously break a local (gauge) symmetry, which can be contrasted with the global case previously considered. Now the symmetric situation has massless gauge bosons as well as massive scalars; the bosons that would be massless after symmetry breaking are eaten by the gauge bosons, which become massive. The single massive leftover scalar boson is still there: That's the Higgs.

While one might argue about whether it was Anderson, Englert and Brout, or Higgs who first proposed the Higgs *mechanism* by which gauge bosons become massive, Higgs himself has a good claim to the first appearance of the Higgs *boson*, the particle that we are now using as evidence that this is how nature works. (Others might point out—and they have—that the earlier papers could have mentioned the Higgs boson but didn't, because its existence should be obvious once the rest of the work is done.) In a follow-up paper in 1966, Higgs examined the properties of this boson in greater detail. But if his original submission of the paper hadn't been rejected by *Physics Letters*, he may never have drawn attention to the boson at all.

Higgs was well aware of Anderson's paper from 1963. He tends to give Anderson substantial credit, but argues that Anderson didn't go far enough: "Anderson should have done basically the two things that I did. He should have shown the flaw in the Goldstone theorem, and he should have produced a simple relativistic model to show it happened. However, whenever I give a lecture on the so-called Higgs mechanism I start off with Anderson, who really got it right, but nobody understood him."

1964: Guralnik, Hagen, and Kibble

Guralnik, Hagen, and Kibble completed their own paper after—but only just barely—the papers by Englert and Brout, and Higgs had already been published. The GHK paper grew out of long-standing discussions between Guralnik and Hagen, who had been undergraduates together at MIT and wrote their first paper together after Hagen stayed at MIT for graduate school and Guralnik moved up the river to Harvard. Those discussions blossomed after Guralnik took a postdoc at Imperial College in London, where Abdus Salam was a faculty member and spontaneous symmetry breaking was a hot topic. Kibble was also there as a faculty member, and he and Guralnik talked frequently about evading the Goldstone theorem. A visit from Hagen provided the impetus for the trio to write up their results in a paper.

According to later recollections, in October 1964, Hagen and Guralnik "were literally placing the manuscript in the envelope to be sent to *PRL*, [when] Kibble came into the office bearing two papers by Higgs and the one by Englert and Brout." The Englert-Brout paper had been submitted on June 26, 1964, and was published in August; Higgs's two papers were submitted on July 27 and August 31, and appeared in September and October, respectively; the GHK paper was submitted on October 12, and appeared in November. Their immediate reaction was to recognize that these heretofore unsuspected works were relevant, but they didn't feel they quite counted as "being scooped." GHK judged that Englert-Brout and Higgs had successfully addressed the question of how gauge bosons could get mass via spontaneous symmetry breaking, but hadn't confronted head-on the issue of what precisely went wrong with Goldstone's theorem, which was a central concern of the Anglo-American triumvirate. They felt that the Englert-Brout discussion of what happened to the various vibrating fields was somewhat obscure, and Higgs's papers were completely classical, not cast in the language of quantum mechanics.

With that in mind, GHK took their paper out of the envelope and added a reference to the slightly earlier works: "We consider, as our example, a theory which was partially solved by Englert and Brout, and bears some resemblance to the classical theory of Higgs." Because nearly simultaneous invention of ideas is fairly common, a convention has arisen in the physics literature: If another paper comes along before yours is quite done, you include a note at the end that references it, with the explanation, "While this work was being completed, we received a related paper by . . ." GHK neglected to do that explicitly, but nobody doubts that their paper was substantially complete before they ever heard of the competing works. It's sufficiently different, and was submitted so soon after the others appeared, that there's no chance they were simply building on the Englert-Brout and Higgs papers.

Guralnik, Hagen, and Kibble undertake a thoroughly quantum-mechanical treatment of the problem of spontaneous breaking of a gauge symmetry. They focus very carefully on the question of how the assumptions of Goldstone's theorem are sidestepped. They do not, however, get the Higgs boson quite right. While the real Higgs is expected to be massive, GHK set its mass to zero by choice. Their explicit statement about this particle is simply, "While one sees by inspection that there is a massless particle in the theory, it is easily seen that it is completely decoupled from the other (massive) excitations, and has nothing to do with the Goldstone theorem." Those statements are true in the model they consider, but only because they set the couplings and mass to zero by hand; in the real world, we expect the Higgs to be massive and coupled to other particles.

There was yet another team pushing in the same direction, although slightly later (by a few months). At the time, scientific communication between the Soviet Union and the West was hampered by numerous bureaucratic restrictions. So in 1965, when physicists Alexander Migdal and Alexander Polyakov—both nineteen years old at the time—were thinking about spontaneous symmetry breaking in gauge theories, they weren't aware of any of the 1964 papers. Their independent work had to suffer a thicket of skeptical reviewers, and didn't appear until 1966.

Despite all this simultaneous activity, many physicists remained skeptical that local symmetries offered a way to escape from the massless particles. Higgs tells the story of giving a seminar at Harvard, where theorist Sidney Coleman had primed his students to "tear apart this joker who thinks he can outsmart Goldstone's theorem." (I can vouch for the veracity of this story, as Coleman related it himself when I took his class on quantum field theory many years later.) But Englert, Brout, Higgs, Guralnik, Hagen, and Kibble had one important fact on their side: They were right. Very soon, their ideas would be put to use in one of the major triumphs that are now incorporated into the Standard Model.

The weak interactions

All of this discussion of different kinds of spontaneous symmetry breaking was concerned with basic questions within quantum field theory: What can happen, and under what circumstances? It remained to be seen whether the phenomena that were described are actually relevant to the real world. It wasn't long, however, before they found a permanent home in our understanding of the weak interactions.

The first promising theory of the weak interactions was invented by Enrico Fermi in 1934. Fermi took advantage of the new idea of the neutrino, which had recently been proposed by Wolfgang Pauli, to develop a model of neutron decay, which we would now say is mediated by the weak interactions. Fermi's calculation was also an early success of quantum field theory, as we discussed in Chapter Seven.

Fermi's theory provided a good fit to the data, but only if you didn't push it too hard. Many calculations in quantum field theory proceed by first finding an approximate answer, and then improving that answer bit by bit, essentially by including the contributions from more complicated Feynman diagrams. In the Fermi theory, the original approximation does a very good job, but the next contribution (which is supposed to be a small correction) turns out to be infinitely big. That's a problem—a big one, which would loom over particle physics throughout the twentieth century. Infinite answers are certainly not right, so they are a sign that your theory is not very good. A theory needs to fit the data, but it also needs to make mathematical sense.

The problem of infinite answers wasn't confined to the weak interactions; it even plagued electromagnetism, which should be one of the simplest and easiest-to-understand quantum field theories there is. There, however, it turns out that the infinities can be tamed. The process for doing so is known as "renormalization," and it's what won the Nobel for Feynman, Schwinger, and Tomonaga.

Some field theories are renormalizable—there are well-defined mathematical techniques for getting finite answers—and some are not. In modern quantum field theory, when a theory fails to be renormalizable, we don't simply throw it away. We just admit that it's an approximation at best, perhaps valid only at very low energies, and that some new physics must be present up at high energies to tame the infinities. For a long time, however, nonrenormalizability was taken as a sign that a theory was simply sick. Fermi's theory of the weak interactions turns out to be nonrenormalizable; it gives infinite answers when we press too hard, and there's no way to fix them beyond coming up with a better theory.

Julian Schwinger, who had been intrigued by the Yang-Mills idea that more elaborate symmetries could produce connection fields that

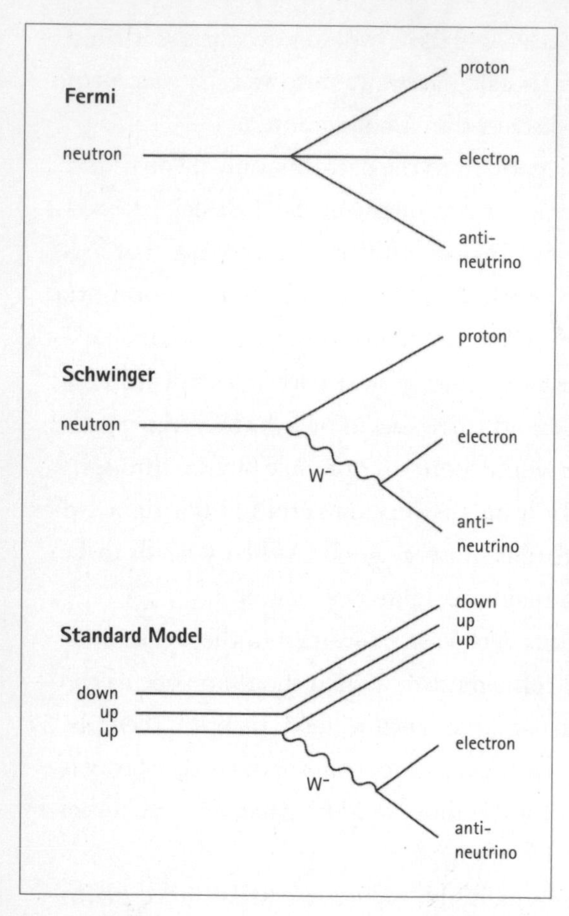

Changing views of the weak interactions, as exemplified by neutron decay. In Fermi's theory, a neutron decays directly to a proton, an electron, and an antineutrino. Schwinger suggested that a charged W^- boson was emitted by the neutron, and then decayed into an electron and an antineutrino. He was right, but we now know the neutron is made of three quarks, one of which changes from a down to an up by emitting a W^-.

accounted for nature's forces, tried to apply the idea to the weak interactions. There is an immediate problem, of course: The Yang-Mills bosons are supposed to be massless, implying a long-range force, while the weak interaction is clearly short-range. Schwinger simply put that problem aside: He started with a Yang-Mills model and made two of the force-carrying bosons massive by hand. This was the first appearance of what we now know as the W^+ and W^- bosons. (One of the first, anyway. In Leon Lederman's words: "Later versions of the Fermi theory, most notably by Schwinger, introduced the heavy W^+ and W^- as weak-force carriers. So did several other theorists. Let's see: Lee, Yang, Gell-Mann . . . I hate to credit any theorists because 99 percent of them will be upset.")

The reason why the Yang-Mills bosons were massless in the first place was because of the symmetry on which the theory was based. When Schwinger gave mass to the bosons it implied that this symmetry was broken, but in this case it was an *explicit* breaking, not a *spontaneous* breaking in which the symmetry was hidden by some field that was nonzero in empty space (which hadn't been invented yet). It wasn't broken because of a field, it was broken because Schwinger said so. As you might guess, this somewhat ad hoc construction wreaked havoc with the model. For one thing, the renormalizability of electromagnetism depends crucially on the symmetry underlying the theory, and disregarding that symmetry rendered Schwinger's model nonrenormalizable. Eventually it was realized that a theory of massive gauge bosons would be renormalizable if and only if the masses came from spontaneous symmetry breaking; but that was years down the road.

Nevertheless, Schwinger didn't persevere with a dodgy theory just because he was stubborn. One property of genius is that you can recognize which kinds of ideas are worth pursuing even though they don't seem to be working quite yet. A nice property of Schwinger's model is that it actually predicted three gauge bosons: the two charged W bosons, which were given a mass, and a single neutral gauge boson, which was allowed to remain massless. We all know about a neutral massless gauge boson, of course: It's the photon. Schwinger was encouraged by the notion that this approach held out the promise of unifying electromagnetism with the weak interactions, which would represent a major step forward in physics. That's probably what kept him going in the face of the problems with the model.

He didn't keep going for very long. Schwinger's paper came out in 1957, and in that same year it was discovered that the weak interactions violate parity. Remember from Chapter Eight (and Appendix One) that particles are either left-handed or right-handed, depending on how they are spinning. Parity violation implied that the weak interactions couple to left-handed particles but not right-handed ones. It's possible to invent Yang-Mills symmetries that involve only left-handed particles, but we know that electromagnetism *doesn't* violate parity—it treats left and right on equal footing. This discovery seemed to put the kibosh on Schwinger's hope of unifying the weak and electromagnetic forces.

Electroweak unification

Sometimes, as a professor, the thing to do is to not give up; it's to hand off your questions to a graduate student. Happily, Schwinger had a very talented young student available: Sheldon Glashow, who was given the task of thinking about unifying electromagnetism and the weak interactions. Glashow has an expansive and charismatic personality, and as a physicist he enjoys hopping quickly from idea to idea. This propensity served him well in the quest for unification, as he was always very willing to propose one theory and then move on quickly to the next one. After thinking about the question on and off for a few years, he hit on a promising scheme for what would ultimately be called "electroweak unification."

The sticking point was parity: Electromagnetism preserves it, while the weak interactions violate it. How could they be unified? Glashow's idea was to introduce two different symmetries: one that treated left-handed and right-handed particles the same, and one that treated them differently. Now, you might think this isn't a step forward; having two different symmetries doesn't sound very unified at all. The secret in Glashow's model was that both symmetries were broken, but in just such a way that a certain combination of the two was left unbroken.

Think of two gear wheels. Either of them can rotate independently; that's like Glashow's original two symmetries. But bring them together, the teeth of both wheels meshing with each other. Now they can still move, but they must move together rather than separately. There is less freedom than before. In Glashow's model, the unbroken symmetry is like the ability to move the wheels together, while the broken symmetry is like the inability to move them at different speeds. The massless, neutral gauge boson corresponding to Glashow's unbroken symmetry is of course the photon.

This idea seemed to be able to accommodate the known features of both the weak and electromagnetic interactions. (It still suffered from the problem that the gauge boson masses were just put in by hand, and the theory wasn't renormalizable.) But it deviated from what was known by predicting a new gauge boson: something that was neutral but massive, what we now call the Z. There was no evidence for such a particle at the time, so the model didn't capture many people's attention.

While the ingredients Glashow put together in his attempt to unify electromagnetism with the weak interactions might seem a bit arbitrary, there was clearly something sensible about them: Across the ocean in Britain, at Imperial College London, almost exactly the same theory was being put together by Abdus Salam and John Ward. Each physicist individually was very accomplished. Ward, who was born in Britain but spent various years living in Australia and the United States, was a pioneer of quantum electrodynamics. He is probably best known within physics for the "Ward identities" in quantum field theory, mathematical relations that enforce local symmetries. Salam, who was born in Pakistan when it was still joined with India under British control, would eventually become politically active and serve as an advocate for science in the developing world. They were frequent collaborators, and some of their most interesting work was done together, on the question of unifying the forces.

Following very similar logic as Glashow's, Salam and Ward invented a model with two different symmetries, one of which violated parity and the other which did not, and which predicted a massless photon and three massive weak gauge bosons. Their paper was published in 1964, apparently without being aware of Glashow's earlier work. Like Glashow, they broke symmetries by hand in their model. Unlike Glashow, they had no excuse for doing so: They were working literally down the hall from Guralnik, Hagen, and Kibble, who were concentrating full-time on spontaneous symmetry breaking.

Part of the failure of communication might have been due to Ward's naturally reticent nature. In his book *The Infinity Puzzle,* Frank Close relates a revealing story told by Gerald Guralnik:

> Guralnik and Ward were having lunch together in a local pub, and Guralnik started to talk about his work—yet to be completed—on hidden symmetry. "I did not get far before [Ward] stopped me. He proceeded to give me a lecture on how I should not be free with my unpublished ideas, because they would be stolen, and often published before I had a chance to finish working on them." As a result of this admonishment, Guralnik did not ask Ward about the work that he himself was doing with Salam.

Even if one takes such a cautious approach to discussing unpublished work, the most secretive physicist usually isn't reluctant to talk about *published* results. For whatever reason, however, Salam and Ward didn't catch on to what Guralnik, Hagen, and Kibble had proposed until several years later. Eventually Salam learned of the work through conversations with Tom Kibble, and for years thereafter would refer to it as the "Higgs-Kibble mechanism."

Putting it all together

The final pieces of the puzzle were put together in 1967. Steven Weinberg had been high school classmates with Sheldon Glashow at the Bronx High School of Science, but they never directly collaborated on the work in theoretical physics that would lead to them sharing the Nobel Prize with Salam in 1977. Today Weinberg is a respected elder statesman of physics, the author of several influential books as well as frequent essays in *The New York Review of Books* and elsewhere. He also was a major advocate for the Superconducting Super Collider—which he would have been even if the accelerator hadn't been located in Texas, where he had moved in 1982.

In 1967 Weinberg was a young professor at MIT, driving a red Camaro to campus each day. He was deeply invested in spontaneous symmetry breaking, but he was using it to try to understand the strong interactions. Inspired by a recent paper by Kibble, Weinberg was playing with a set of symmetries that, unbeknownst to him at the time, bore a close resemblance to those considered by Glashow and Salam and Ward before. The problem was that he kept predicting a massless, neutral, gauge boson, which didn't seem to be there in the strong interactions.

In September of that year, Weinberg suddenly realized that he had been thinking about the wrong problem. His problematic model of the strong interactions worked very well as a theory of the weak and electromagnetic interactions. The annoying massless boson was a feature, not a bug: It was the photon. In a short paper entitled "A Theory of Leptons," Weinberg put together what every modern graduate student in particle physics would immediately recognize as what's known as the "electroweak" sector of the Standard Model. In the references he cited Glashow's paper, but he still wasn't aware of the one by Salam and Ward. Using Kibble's ideas, he was able to make a direct prediction for the masses of the W and Z bosons—something Glashow and Salam and Ward weren't able to do, as they had inserted the masses by hand. Weinberg accounted for the mechanism by which all the fermions in the theory acquired mass, as well as the gauge bosons. He even noted that the model might possibly be renormalizable, although he wasn't able to offer any convincing arguments at the time. A coherent theory of electroweak unification had finally been assembled.

At almost precisely the same time, Kibble and Salam finally realized their mutual interest in symmetry breaking, and Kibble explained the theory to Salam. Salam figured out that he could rework the unified model he had proposed with Ward to include symmetry-breaking scalar bosons, and gave lectures on his ideas to a small audience at Imperial. For unknown reasons, Salam didn't write up these ideas right away; he was extremely prolific as a physicist, but his major focus in those days was on gravity, not on subatomic forces. Consequently, his proposal to add a Higgs mechanism to the Salam-Ward model didn't appear in print until a year later, when he included it in the proceedings from a conference talk (where he also cites Weinberg's paper).

The separate papers by Weinberg and Salam had all the impact, as Kurt Vonnegut once said in a different context, of a pancake twelve feet in diameter dropped from a height of two inches. In academia, and science in particular, the most concrete way of quantifying the influence of a piece of research is to count how many times the paper has been cited by other papers. Between 1967 and 1971, Weinberg's paper was cited just a handful of times. The two authors did not even pursue their own ideas to any great extent in the years immediately thereafter. Since 1971, however, Weinberg's paper has been cited more than 7,500 times—an average of more than once every two days for four decades.

What happened in 1971? Some surprising experimental result? No, a surprising theoretical result: Gerard 't Hooft, a young graduate student in the Netherlands, working under Martinus "Tini" Veltman, proved that theories with spontaneously broken gauge symmetries are renormalizable, even though the gauge bosons are massive. In other words, 't Hooft showed that the electroweak theory made mathematical sense. This had been conjectured by both Weinberg and Salam, but many people in the field had remained skeptical, which partly accounts for the obscurity of these ideas up to that point. In Sidney Coleman's words, 't Hooft "revealed Weinberg and Salam's frog to be an enchanted prince." Gerard 't Hooft has since gone on to earn a reputation as one of the most creative and brilliant minds in physics. He and Veltman shared the Nobel Prize in 1999 for their work on the electroweak theory and spontaneous symmetry breaking.

The surprising experimental results weren't long in coming, however. The main novel prediction of the Glashow, Salam-Ward, and

Weinberg models was the existence of a heavy neutral boson, the Z. The effects of the W bosons were well-known: They change the identity of a fermion when they are emitted (for example, changing a down quark to an up during neutron decay). If the Z existed, it would imply a version of the weak interactions in which particles kept their identities; for example a neutrino could scatter off an atomic nucleus. Events of precisely this kind were observed at CERN's Gargamelle detector in 1973, setting the stage for Glashow, Salam, and Weinberg to share the Nobel Prize in 1979. (Ward was left out, but only three people can share the prize in any one year.) The W and Z bosons themselves, as opposed to their indirect effects, weren't discovered until Carlo Rubbia found them a few years afterward.

All that remained was to discover the Higgs boson.

The name game

Physicists are human beings. They are typically motivated by what Richard Feynman called "the pleasure of finding things out," but once they find out something interesting they appreciate getting credit for their work. Throughout this book, following nearly universal practice within the physics community, I've been referring to the "Higgs mechanism" for given mass to gauge bosons via spontaneous symmetry breaking, as well as the "Higgs boson" for the scalar particle that this model predicts. It's clear, however, that while Higgs's contributions were important, he was hardly alone. Why is that the name, and what should be the name?

Nobody is precisely sure where the name "Higgs boson" originally came from; it certainly wasn't from Higgs himself. Particle physics lore points the finger at Benjamin Lee, a talented Korean-American physicist who died in a tragic car accident in 1977. Lee had learned about spontaneous breakdown of gauge symmetries from talking with Higgs, and the story goes that he gave an influential talk at a conference at

Fermilab in 1972, where he repeatedly referred to the "Higgs meson." That was in the immediate aftermath of 't Hooft's revolutionary result, when everyone was scrambling to learn about these ideas. Precisely because physicists are human beings, they tend to lazily stick with the first words they hear attached to the subject, so a widely heard talk can spread a piece of nomenclature far and wide.

Another theory goes back to Weinberg's 1967 paper. When the original papers came out in 1964, not too many physicists were thinking about spontaneous symmetry breaking in gauge theories; after 't Hooft's breakthrough in 1971, many rushed to catch up, and Weinberg's paper was a good starting point. In his discussion of the Higgs mechanism, he references three papers by Higgs, as well as the paper by Englert and Brout and the one by Hagen, Guralnik, and Kibble. However, Higgs comes first in his reference list, due to a mix-up between *Physical Review Letters* (where Higgs's second paper appeared) and *Physics Letters* (where the paper by Englert and Brout appeared). From such minor lapses are long-lasting consequences forged.

Perhaps most important, "Higgs boson" sounds like a good name for a particle. It was Higgs's papers that first drew close attention to the boson particle rather than the "mechanism" from which it arose, but that's not quite enough to explain the naming convention. We might ask, however, what is the alternative? There may have been a chance, in the early days, to come up with a label that wasn't derived from the name of a person. The "radial boson," perhaps, or the "relicon," since the boson is the only surviving relic of the symmetry-breaking process. The "electroweak boson" would work, although it runs the danger of being confused with the W and Z bosons, so the "electroweak scalar boson" might be most accurate.

But absent such a construction (and it's not as if those suggestions are very good), it's hard to do justice to the history by choosing a naming convention. Higgs himself refers to "the boson that has been named after me," and sometimes talks about the "ABEGHHK'tH mechanism"—that's Anderson, Brout, Englert, Guralnik, Hagen, Higgs, Kibble, and 't Hooft, for those of you scoring at home. Joe Lykken at Fermilab switched out 't Hooft in favor of Nambu to come up with "HEHK-BANG," which is at least a pronounceable acronym, but no more attractive. "That would be foolish," as he himself admits.

Ultimately one has to admit that the name of a particle is just a label. It's not supposed to be, and shouldn't be taken as, a comprehensive and fair history of the development of an idea. We can call it the "Higgs boson" without pretending that Higgs is the only one who deserves credit. (Given the funding pressures in modern particle physics, I suspect that the naming rights would be happily sold for about $10 billion. "The McDonald's boson," anyone?)

The verdict of history

As we have recounted the story, Nambu and Goldstone helped establish our understanding of spontaneous symmetry breaking, but they concentrated on the case of global symmetries. Anderson pointed out that gauge symmetries are different, and in particular that they didn't leave any remnant massless particles, but he didn't construct an explicitly relativistic model. That was done independently by Englert and Brout, by Higgs, and by Guralnik, Hagen, and Kibble. All three took slightly different routes but achieved essentially the close up same answers, and all three deserve a hefty measure of credit. As does 't Hooft, who showed that the idea made mathematical sense.

By tradition, the Nobel Prize in sciences is given to individuals rather than groups, and no more than three individuals in any one year. There's no question that the candidates are jockeying for position, at least discreetly. 't Hooft and Veltman have already won a Nobel for their work on renormalizing electroweak theory. Anderson won a Nobel for something completely different, but realistically that does hurt his chances for a second prize (even if he does have a good case for being there first). Robert Brout passed away in 2011, and Nobels are not given posthumously.

In 2004, the Wolf Prize in Physics—sometimes described as the second-most prestigious award after the Nobel—was given to Englert, Brout, and Higgs, but not to Guralnik, Hagen, and Kibble. At a 2010 "Higgs Hunting" meeting in France, the advertising poster made direct mention of "Brout, Englert, and Higgs," leaving out GHK entirely. This caused a certain amount of push-back, with supporters of the Anglo-American team threatening to boycott the conference. Organizer Gregorio Bernardi was taken aback by the criticism, saying, "People took this very seriously, which we didn't expect." That seems at least somewhat disingenuous; if you care enough about assigning credit that you attach the names of Englert and Brout to a boson that has universally been known as the "Higgs," you can't be surprised when Guralnik, Hagen, and Kibble (or their partisans) are upset. Part of the sting was taken away when the American Physical Society awarded its 2010 Sakurai Prize in theoretical physics to Hagen, Englert, Guralnik, Higgs, Brout, and Kibble—in that order, which seems to have been chosen specifically to make it impossible for anyone to complain. (Anderson might have reasonably complained.)

As Anderson ruefully notes, "If you want the history right in detail, you better write it yourself." Over the past several years Guralnik, Higgs, Kibble, and Brout and Englert have all written reminiscences of their work in 1964, attempting to put their own contributions in perspective. And, this being the modern age, a controversy flared up on *Wikipedia*, the online encyclopedia that can be edited by anyone. In August 2009, a user known only as "Mary at CERN" put up a new entry entitled "1964 PRL Symmetry Breaking Papers." There were already separate entries on "Spontaneous Symmetry Breaking," "Higgs Mechanism," and so forth; this new article aimed squarely at the question of how credit should be attributed. While discussing all the papers, it was clear who the new entry was meant to support: "A case can be made that, while first to publish by a couple months, Higgs and Brout-Englert solved half of the problem—massifying the gauge particle. Guralnik-Hagen-Kibble, while published a couple months later, had a more complete solution—massifying the gauge particle and also showing how the numbing influence from Goldstone's theorem is avoided." But what one person can write on *Wikipedia*, another can edit; the current revision is a bit more even-handed.

I have no particular preference concerning who, if anyone, should win the Nobel Prize for inventing the idea of the Higgs boson, nor do I have a prediction. The prizes are good for science, as they help draw attention to interesting work that might not otherwise be publicized. But they're not what science is about; the reward for helping to discover the mechanism in the first place is enormously larger than any prize the Nobel committee can bestow.

The real disappointment is that it seems difficult to imagine any experimentalist claiming a Nobel for actually discovering the boson. It's a simple problem of numbers: Too many people contributed to the experiments in too many ways for any one or two or three to be picked out as responsible. One achievement that is unquestionably Nobel-worthy is the successful construction of the LHC itself, so Lyn Evans would be a sensible candidate. It's probably past time for the Nobel foundation to think about relaxing the tradition that collaborations cannot win any of the prizes in science. Whoever gets that rule change implemented might deserve the Nobel Peace Prize.

TWELVE

BEYOND THIS HORIZON

In which we consider what lies beyond the Higgs boson: worlds of new forces, symmetries, and dimensions?

From the age of ten, Vera Rubin was fascinated by the stars. Her interest never waned, and when she applied to college it was natural that she would seek to study astronomy. But this was in the 1940s, and women were not exactly welcome in science. At one point she spoke to a Swarthmore College admissions officer, who asked whether she had any other interests. She admitted that she enjoyed painting. The admissions officer seized on that, asking, "Have you ever considered a career in which you paint pictures of astronomical objects?" She ended up attending Vassar College instead, but the question made an impression. She later recalled, "That became a tag line in my family: for many years, whenever anything went wrong for anyone, we said, 'Have you ever considered a career in which you paint pictures of astronomical objects?' "

Rubin persevered, proceeding to graduate studies at Cornell and Georgetown University. The road wasn't easy; when she wrote to Princeton asking for a graduate school catalogue, they refused to send her one, noting that the astronomy department didn't accept female graduate students. (That policy eventually ended in 1975.)

One secret to success as a scientist is to look where others don't. As larger telescopes were becoming available, many astronomers turned their gaze to the centers of distant galaxies, in regions rich with stars and activity. Rubin chose to concentrate on their outer fringes, studying the dynamics of the thinly spread stars and gas orbiting slowly on the edges. This technique provides a way to measure the total mass of a galaxy: The more matter inside, the higher the gravitational field on the outer stars will be, and the faster they will have to orbit.

Rubin and her collaborator Kent Ford found something astonishing. We expect that stars should move more and more slowly as we move away from the center of the galaxy, just as more distant planets in the solar system orbit more slowly around the sun. The gravitational field is lower, so there is less force to resist, requiring less velocity to maintain an orbit. But Rubin and Ford found something very different: Stars move at equal speeds as we examine larger and larger distances from the dense central region of a galaxy. The implication is straightforward, although hard to accept: There is much more matter in a galaxy than we observe, and much of it is distributed far from the center, unlike the visible stars.

What Rubin and Ford had stumbled upon was a surprising phenomenon that today sits at the center of modern cosmology: dark matter.

They weren't the first; as far back as the 1930s, Swiss-American astronomer Fritz Zwicky had demonstrated that there was much more matter in the Coma cluster of galaxies than we can observe in our telescopes, and Dutch astronomer Jan Oort showed that our local galactic neighborhood had more matter than was immediately evident. For a long time, however, there was hope that the matter was simply "missing"; it was just ordinary stuff, but in a form that wasn't easy to see. As we learned more and more about galaxies and clusters and the universe as a whole, we were able to precisely measure two numbers separately: the total amount of matter in the universe and the total amount of "ordinary matter," where ordinary matter includes atoms, dust, stars, planets, and every kind of particle known in the Standard Model.

The two numbers don't match. The total amount of ordinary

matter in the universe is only about one-fifth of the total amount of matter. The vast majority is dark matter, and the dark matter can't be any of the particles in the Standard Model.

The Higgs boson is the final piece of the Standard Model puzzle, but the Standard Model is certainly not the end of the road. Dark matter is just one indication that there is a lot more physics out there remaining to be understood. One exciting prospect is that the Higgs can serve as a bridge between what we know and what we hope to learn. By studying carefully the properties of the Higgs, we hope to shed light on the dark worlds beyond our own.

The early universe

Let's think about dark matter a bit more carefully, as it provides some of the strongest evidence we have for physics beyond the Standard Model, and a great example of how the Higgs could be involved in understanding this new physics better. A crucial feature of dark matter is that it can't be ordinary matter (atoms and so forth) in some "dark" form, like brown dwarfs or planets or interstellar dust. That's because we have very good measurements of the total amount of ordinary matter, from processes in the early universe.

To understand dark matter, we need to think about where it came from. Imagine you have an experimental apparatus that is basically a super-oven: a sealed box with some stuff inside, and a dial you can adjust to make the temperature as high or low as you like. An ordinary oven might go as high as 500 degrees Fahrenheit, which in particle physics units is about 0.04 electron volts. At that temperature, molecules can rearrange themselves (which we call "cooking"), but atoms maintain their integrity. Once we get up to a few electron volts or higher, electrons are stripped away from their nuclei. When we hit millions of electron volts (MeV), the nuclei themselves are stripped apart, leaving us with free protons and neutrons.

Another thing also happens at high temperatures: The collisions between particles are so energetic that you can create new particle-antiparticle pairs, just like in a particle collider. If the temperature is higher than the total mass of a particle plus its antiparticle, we expect such pairs to be copiously produced. So at sufficiently high temperatures, it almost doesn't matter what was in the box to begin with; what we get is a hot plasma filled with all the particles that have masses lower than the temperature inside. (Remember that mass and temperature can both be measured in GeV.) If the temperature is 500 GeV, our box will be buzzing with Higgs bosons, all the quarks and leptons, W and Z bosons, and so forth—not to mention possible new particles that haven't yet been discovered here on earth. If we were to gradually lower the temperature inside of that box, these new particles would gradually disappear as they bumped into their antiparticles and annihilated, leaving us with only the particles we started with.

The early universe is much like the plasma inside our ultrahot oven, with one crucial extra ingredient: Space is expanding at an incredible rate. The expansion of space has two important effects. First, the temperature cools off, so it's as if the temperature dial on our oven starts very high but is quickly turned down. Second, the density of matter decreases rapidly as particles move away from one another in the expanding space. That latter feature is a crucial difference between the early universe and an oven. Because the density is decreasing, particles that were produced in the original plasma might not have a chance to annihilate away; it might simply be too hard for one of them to find a corresponding antiparticle.

As a result, we get a relic abundance of these particles from the primordial plasma. And we can calculate precisely what that abundance should be, if we know the masses of the particles and the rate at which they interact. If the particles are unstable, like the Higgs boson is, the relic abundance is pretty irrelevant, as the particles just decay away. But if they're stable, we're stuck with them. It's easy to imagine that a leftover stable particle from the early universe constitutes the dark matter today.

In the Standard Model, we can play this game with the atomic nuclei. One crucial difference is that we start with more matter than antimatter, so the matter can never completely annihilate away. Start at a fairly high temperature, say around 1 GeV. The plasma will consist of protons, neutrons, electrons, photons, and neutrinos; all the heavier particles will have decayed. That temperature is sufficiently hot that protons and neutrons cannot form nuclei without being instantly ripped apart. But as the universe expands and cools, nuclei begin to form a few seconds after the Big Bang. Just a couple of minutes later, the density is so low that nuclei stop running into one another, and those reactions cease. We are left with a certain combination of protons and light elements: deuterium (heavy hydrogen, one proton and one neutron), helium, and lithium. This process is known as "Big Bang nucleosynthesis."

We can make precise calculations of the relative abundance of those elements, with just one input parameter: the initial abundance of protons and neutrons. And then we can compare the primordial element abundances with what we see in the real universe. The answer matches precisely, but only for one specific density of protons and neutrons. That happy result is reassuring, since it indicates that the way we think about the early universe is basically on the right track. Since protons and neutrons make up the overwhelming fraction of mass in ordinary matter, we know quite well how much ordinary matter there is in the universe, no matter what form it might take today. And it's not nearly enough to make up all the matter there is.

WIMPs

One promising strategy for dark matter is to play that same game as we did with nucleosynthesis, but starting at a much higher temperature and adding a new particle into the mix—a particle that will be the dark matter. We know that dark matter is dark, so the new particle should be electrically neutral. (Charged particles are precisely those that interact with electromagnetism, and therefore tend to give off light.) And we know it's still around, so it should be stable, or at least have a lifetime longer than the age of the universe. We even know something a bit more detailed: Dark matter doesn't interact very strongly with itself. If it did, it would settle into the middle of galaxies, rather than forming big puffy halos as the data seem to indicate. So the dark matter doesn't feel the strong nuclear force, either. Of the known forces of nature, the dark matter certainly feels gravity, and it may or may not feel the weak nuclear force.

Let's imagine a particular kind of new particle: a "Weakly Interacting Massive Particle," or WIMP. (Cosmologists are nothing if not cheeky when it comes to inventing new names.) By "weakly interacting" we don't just mean "doesn't interact very much"; we mean that it feels the weak interactions of particle physics. For simplicity, we assume the WIMP has a mass compatible with other particles involved in the weak interactions, like the W and Z bosons or the Higgs. Around 100 GeV, let's say, or at least between 10 and 1,000 GeV. Other details of the way such a particle interacts are relevant for high-precision calculations, but just these basic properties are enough to perform back-of-the-envelope estimates.

Then we compare the predicted abundance of such a WIMP with the actual abundance of dark matter. What we find—amazingly—is that they match beautifully. There's some wiggle room, having to do with what other particles might exist and how exactly the WIMPs annihilate, but the rough agreement is striking. Stable particles with weak-scale interactions generally have the right relic abundance to account for the dark matter, without even trying too hard.

This interesting coincidence is known as the "WIMP miracle" and has given many particle physicists hope that the secret to dark matter lies in new particles with similar masses and interactions to the W/Z/Higgs bosons. All those particles decay quickly, of course, so the WIMP must have some good reason to be stable, but that's not hard to invent. There are many other plausible theories of dark matter—including a particle called the "axion," invented by Steven Weinberg and Frank Wilczek, which is like a very lightweight cousin of the Higgs—but WIMP models are by far the most popular.

The possibility that the dark matter is a WIMP opens up some very exciting experimental possibilities, precisely because the Higgs will interact with it. Indeed, in many (arguably most, but it's hard to count) viable models of WIMP dark matter, the strongest coupling between the dark matter and ordinary matter will be through exchanging a Higgs boson. The Higgs could be the link between our world and most of the matter in the universe.

The Higgs portal

This feature—interacting via Higgs exchange—turns out to be common in many theories of physics beyond the Standard Model. You have a whole bunch of new particles in what's known as a "hidden sector," and they don't interact very noticeably with the particles we've already studied. The Higgs is a little more sociable than the known fermions and gauge bosons, which means that it's more likely to interact with the new particles. That's the sense in which our discovery of the Higgs is both the completion of one grand project—constructing the Standard Model—but also the beginning of the next—finding hidden worlds beyond that model. Wilczek and his collaborator Brian Patt have dubbed this possibility the "Higgs portal" between the Standard Model and hidden sectors of matter.

In discussing Higgs detection in Chapter Nine, I drew attention to the decay of the Higgs into two photons, which was mediated by a loop of virtual particles. The actual rate at which such a process occurs depends on all the different particles that can possibly appear in that loop—that is, particles that couple both to the Higgs and to photons. In the Standard Model itself, this rate is completely fixed once we know the Higgs mass. Therefore, if we carefully measure this decay and find that it proceeds more rapidly than we expect, that serves as strong evidence for the existence of new particles, even if we don't see them directly. The LHC data from 2011 and early 2012 seemed to indicate that more photons were being produced than the Standard Model predicts, even though the difference was not extremely significant. That's certainly something to be watching for as more data are collected.

In the WIMP scenario, dark matter is all around us, even right where you're sitting this moment. In our local environment, we expect there to be roughly one dark-matter particle per coffee-cup-size volume of space. But the particles are moving quite rapidly, typically at hundreds of kilometers per second. As a result, billions of WIMPs pass through your body every second. Because they interact so weakly, you hardly notice; most WIMPs literally go right through you without ever interacting. But although the interactions are small, they're not quite zero. By exchanging a Higgs boson, a WIMP can bump into one of the quarks contained in the protons and neutrons inside your body. Physicists Katherine Freese and Christopher Savage have calculated that in reasonable models, we expect about ten dark-matter particles to interact with the atoms in a typical human body every year. The effects of every individual interaction are pretty negligible, so don't worry about getting a dark matter stomachache.

We can, however, use this kind of interaction to search for dark matter. Just like in the LHC, a crucial task is to separate signals from background noise. Dark matter isn't the only thing that can bump into a nucleus; radioactivity and cosmic rays do it all the time. Physicists therefore go deep underground, into mine shafts and specially built facilities, where they are as shielded from these pesky backgrounds as possible. They then build detectors that patiently wait for the faint signal of a dark-matter particle passing through and perturbing a nucleus. Two types of detectors are popular: cryogenic, where the detector registers the heat created when a dark-matter particle collides with a nucleus inside a low-temperature crystal, and liquid noble gas, where

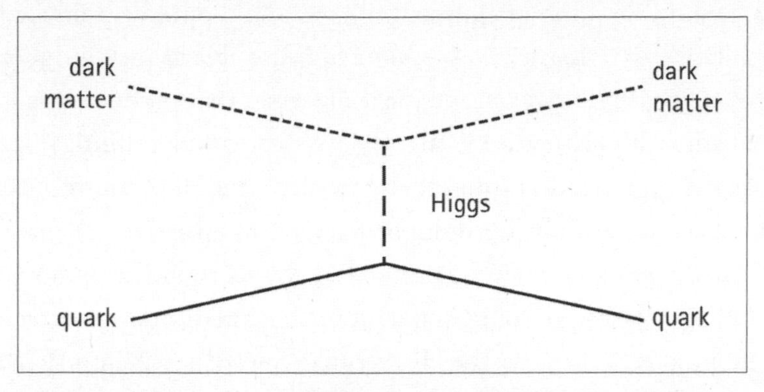

A Feynman diagram representing a dark-matter particle scattering off a quark by exchanging a Higgs boson.

the detector measures light produced through scintillation when a dark-matter particle interacts with liquid xenon or argon.

The strategy of going deep underground and searching for interactions with ambient dark-matter particles is known as "direct detection," and is an ongoing high-priority research frontier. A number of experiments have already ruled out some of the possible models. Knowing the mass of the Higgs boson will help relate the theoretical predictions for WIMP properties to the possible signatures these experiments might see. With the sensitivity already impressive and rapidly improving, nobody should be surprised if we finally detect dark matter once and for all sometime in the next five years. Nor should anyone be surprised if we don't; nature always has surprises.

Naturally, if there is a technique called "direct detection," there is a different technique called "indirect detection." The idea here is to wait for WIMPs in our galaxy or others to collide with one another and annihilate. Among the particles produced in such an interaction will be gamma rays (high-energy photons), which can be searched for using satellite observatories. Currently, NASA's Fermi Gamma-ray Space Telescope is scanning the sky, observing gamma rays, and building up a database of high-energy phenomena. Once again, the problem of separating signal from noise is severe. Astronomers are working hard to understand what kind of gamma-ray signature should be produced by annihilating dark matter, in the hope of being able to pick it out from the many conventional astrophysical processes that produce this kind of radiation. It's also possible that dark matter could annihilate into a Higgs boson (instead of into other particles via a Higgs boson), a scenario that has naturally been dubbed "Higgs in Space."

Finally, we can imagine creating dark matter right here at home, at the LHC. If the Higgs couples to dark matter, and the dark matter isn't too heavy, one of the ways the Higgs can decay is directly into WIMPs. We can't detect the WIMPs, of course, since they interact so weakly; any that are produced will fly right out of the detector, just as neutrinos do. But we can add up the total number of observed Higgs decays and compare it with the number we expect. If we're getting fewer than anticipated, that might mean that some of the time the Higgs is decaying into invisible particles. Figuring out what such particles are, of course, might take some time.

An unnatural universe

Dark matter is solid evidence that we need physics beyond the Standard Model. It's a straightforward disagreement between theory and experiment, the kind physicists are used to dealing with. There's also a different kind of evidence that new physics is needed: fine-tuning within the Standard Model itself.

To specify a theory like the Standard Model, you have to give a list of the fields involved (quarks, leptons, gauge bosons, Higgs), but also the values of the various numbers that serve as parameters of the theory. These include the masses of the particles as well as the strength of each interaction. The strength of the electromagnetic interaction, for example, is fixed by a number called the "fine-structure constant," a famous quantity in physics that is numerically close to 1/137. In the early days of the twentieth century, some physicists tried to come up with clever numerological formulas to explain the value of the fine-structure constant. These days, we accept that it is simply part of the input of the Standard Model, although there is still hope that a more unified theory of the fundamental interactions would allow us to calculate it from first principles.

Although all of these numbers are quantities we have to go out and measure, physicists still believe that there are "natural" values for them to take on. That's because, as a consequence of quantum field theory, the numbers we measure represent complicated combinations of many different processes. Essentially, we need to add up different contributions from various kinds of virtual particles to get the final answer. When we measure the charge of the electron by bouncing a photon off it, it's not just the electron that is involved; that electron is a vibration in a field, which is surrounded by quantum fluctuations in all sorts of other fields, all of which add up to give us what we perceive as the "physical electron." Each arrangement of virtual particles contributes a specific amount to the final answer, and sometimes the amount can be quite large.

It would be a big surprise, therefore, if the observed value of some quantity was very much smaller than the individual contributions that went into creating it; that would mean that large positive contributions had added to large negative contributions to create a tiny final result. It's possible, certainly; but it's not what we would anticipate. If we measure a parameter to be much smaller than we expect, we declare there is a fine-tuning problem, and we say that the theory is "unnatural." Ultimately, of course, nature decides what's natural, not us. But if our theory seems unnatural, it's a sign that we might need a better theory.

For the most part, the parameters of the Standard Model are pretty natural. There are two glaring exceptions: the value of the Higgs field in empty space, and the energy density of empty space, also known as the "vacuum energy." Both are much smaller than they have any right to be. Notice that they both have to do with the properties of empty space, or the "vacuum." That's an interesting fact, but not one that anyone has been able to take advantage of to solve the problems.

The two problems are very similar. For both the value of the Higgs field and the vacuum energy, you can start by specifying any value you like, and then on top of that you imagine calculating extra contributions from the effects of virtual particles. In both cases, the result is to make the final result bigger and bigger. In the case of the Higgs field value, a rough estimate reveals that it should be about 10^{16}—ten quadrillion—times larger than it actually is. To be honest, we can't be too precise about what it "should be," since we don't have a unified theory of all interactions. Our estimate comes from the fact that virtual particles want to keep making the Higgs field value larger, and there's a limit to how high it can reasonably go, called the "Planck scale." That's the energy, about 10^{18} GeV, where quantum gravity becomes important and spacetime itself ceases to have any definite meaning.

This giant difference between the expected value of the Higgs field in empty space and its observed value is known as the "hierarchy problem." The energy scale that characterizes the weak interactions (the Higgs field value, 246 GeV) and the one that characterizes gravity (the Planck scale, 10^{18} GeV) are extremely different numbers; that's the hierarchy we're referring to. This would be weird enough on its own, but we need to remember that quantum-mechanical effects of virtual particles want to drive the weak scale up to the Planck scale. Why are they so different?

Vacuum energy

As if the hierarchy problem weren't bad enough, the vacuum energy problem is much worse. In 1998, astronomers studying the velocities of distant galaxies made an astonishing discovery: The universe is not just expanding, it's accelerating. Galaxies aren't just moving away from us, they're moving away faster and faster. There are different possible explanations for this phenomenon, but there is a simple one that is currently an excellent fit to the data: vacuum energy, introduced in 1917 by Einstein as the "cosmological constant."

The idea of vacuum energy is that there is a constant of nature that tells us how much energy a fixed volume of completely empty space will contain. If the answer is not zero—and there's no reason it should be—that energy acts to push the universe apart, leading to cosmic acceleration. The discovery that this was happening resulted in a Nobel Prize in 2011 for Saul Perlmutter, Adam Riess, and Brian Schmidt.

Brian was my office mate in graduate school. In my last book, *From Eternity to Here*, I told the story of a bet he and I had made back in those good old days; he guessed that we would not know the total density of stuff in the universe within twenty years, while I was sure that we would. In part due to his own efforts, we are now convinced that we do know the density of the universe, and I collected my winnings—a small bottle of vintage port—in a touching ceremony on the roof of Quincy House at Harvard in 2005. Subsequently, Brian, who is a world-class astronomer but a relentlessly pessimistic prognosticator, bet me that we would fail to discover the Higgs boson at the LHC. He recently conceded that bet as well. Both of us having grown, the stakes have risen accordingly; the price of Brian's defeat is that he is using his frequent-flyer miles to fly me and my wife, Jennifer, to visit him in Australia. At least Brian is in good company; Stephen Hawking had a $100 bet with Gordon Kane that the Higgs wouldn't be found, and he's also agreed to pay up.

To explain the astronomers' observations, we don't need very much vacuum energy; only about one ten-thousandth of an electron volt per cubic centimeter. Just as we did for the Higgs field value, we can also perform a back-of-the-envelope estimate of how big the vacuum energy should be. The answer is about 10^{116} electron volts per cubic centimeter. That's larger than the observed value by a factor of 10^{120}, a number so big we haven't even tried to invent a word for it. The hierarchy problem is bad, but the vacuum energy problem is numerically much worse.

Understanding the vacuum energy is one of the leading unsolved problems of contemporary physics. One of the many contributions that makes the estimated vacuum energy so large is that the Higgs field, sitting there with a nonzero value in empty space, should carry a lot of energy (positive or negative). This was one of the reasons Phil Anderson was wary of what we now call the Higgs mechanism; the large energy density of a field in empty space seems to be incompatible with the relatively small energy density that empty space actually has. Today we don't think of this as a deal breaker for the Higgs mechanism, simply because there are also plenty of other contributions to the vacuum energy that are even larger, so the problem runs much deeper than the Higgs contribution.

It's also possible that the vacuum energy is exactly zero, and that the universe is being pushed apart by a form of energy that is slowly decaying rather than strictly constant. That idea goes under the rubric of "dark energy," and astronomers are doing everything they can to test whether it might be true. The most popular model for dark energy is to have some new form of scalar field, much like the Higgs, but with an incredibly smaller mass. That field would roll gradually to zero energy, but it might take billions of years to do it. In the meantime, its energy would behave like dark energy—smoothly distributed through space, and slowly varying with time.

The Higgs boson we've detected at the LHC isn't related to vacuum energy in any direct way, but there are indirect connections. If we knew more about it, we might better understand why the vacuum energy is so small, or how there could be a slowly varying component of dark energy. It's a long shot, but with a problem this stubborn we have to take long shots seriously.

Supersymmetry

A major lesson of the success of the electroweak theory is that symmetry is our friend. Physicists have become enamored with finding as much symmetry as they possibly can. Perhaps the most ambitious attempt along these lines comes with a name that is appropriate if not especially original: supersymmetry.

The symmetries underlying the forces of the Standard Model all relate very similar-looking particles to one another. The symmetry of the strong interactions relates quarks of different colors, while the symmetry of the weak interactions relates up quarks to down quarks, electrons to electron neutrinos, and likewise for the other pairs of fermions. Supersymmetry, by contrast, takes the ambitious step of relating fermions to bosons. If a symmetry between electrons and electron neutrinos is like comparing apples to oranges, trying to connect fermions with bosons is like comparing bananas to orangutans.

At first glance, such a scheme doesn't seem very promising. To say that there is a symmetry is to say that a certain distinction doesn't matter; we label quarks "red," "green," and "blue," but it doesn't matter which color is which. Electrons and electron neutrinos are certainly different, but that's because the weak interaction symmetry is broken by the Higgs field lurking in empty space. If the Higgs weren't there, the (left-handed parts of the) electron and electron neutrino would indeed be indistinguishable.

When we look at the fermions and bosons of the Standard Model, they look completely unrelated. The masses are different, the charges are different, the way certain particles feel the weak and strong forces and others don't is different, even the total number of particles is completely different. There's no obvious symmetry hiding in there.

Physicists tend to persevere, however, and eventually they hit on the idea that all the particles of the Standard Model have completely new "superpartners," to which they are related by supersymmetry. All of these superpartners are supposed to be very heavy, so we haven't detected any of them yet. To celebrate this clever idea, physicists invented a cute naming convention. If you have a fermion, tag "s–" onto the beginning to label its boson superpartner; if you have a boson, tag "–ino" onto the end to label its fermion superpartner.

In supersymmetry, therefore, we have a new set of bosons called "selectrons" and "squarks" and so on, as well as a new set of fermions called "photinos" and "gluinos" and "higgsinos." (As Dave Barry likes to say, I swear I am not making this up.) The superpartners have the same general features as the original particles, except the mass is much larger and bosons and fermions have been interchanged. Thus, a "stop" is the bosonic partner of the top quark; it feels both the strong and weak interactions, and has charge +2/3. Interestingly, in specific models of supersymmetry the stop is often the lightest bosonic superpartner, even though the top is the heaviest fermion. The fermionic superpartners tend to mix together, so the partners of the W bosons and the charged Higgs bosons mix together to make "charginos," while the partners of the Z, the photon, and the neutral Higgs bosons mix together to make "neutralinos."

Supersymmetry is currently an utterly speculative idea. It has very nice properties, but we don't have any direct evidence in its favor. Nevertheless, the properties are sufficiently nice that it has become physicists' most popular choice for particle physics beyond the Standard Model. Unfortunately, while the underlying idea is very simple and elegant, it is clear that supersymmetry must be broken in the real world; otherwise particles and superpartners would have equal masses. Once we break supersymmetry, it goes from being simple and elegant to being a god-awful mess.

There is something called the "Minimal Supersymmetric Standard Model," which is arguably the least complicated way to add supersymmetry to the real world; it comes with 120 new parameters that must be specified by hand. That means that there is a huge amount of freedom in constructing specific supersymmetric models. Often, to make

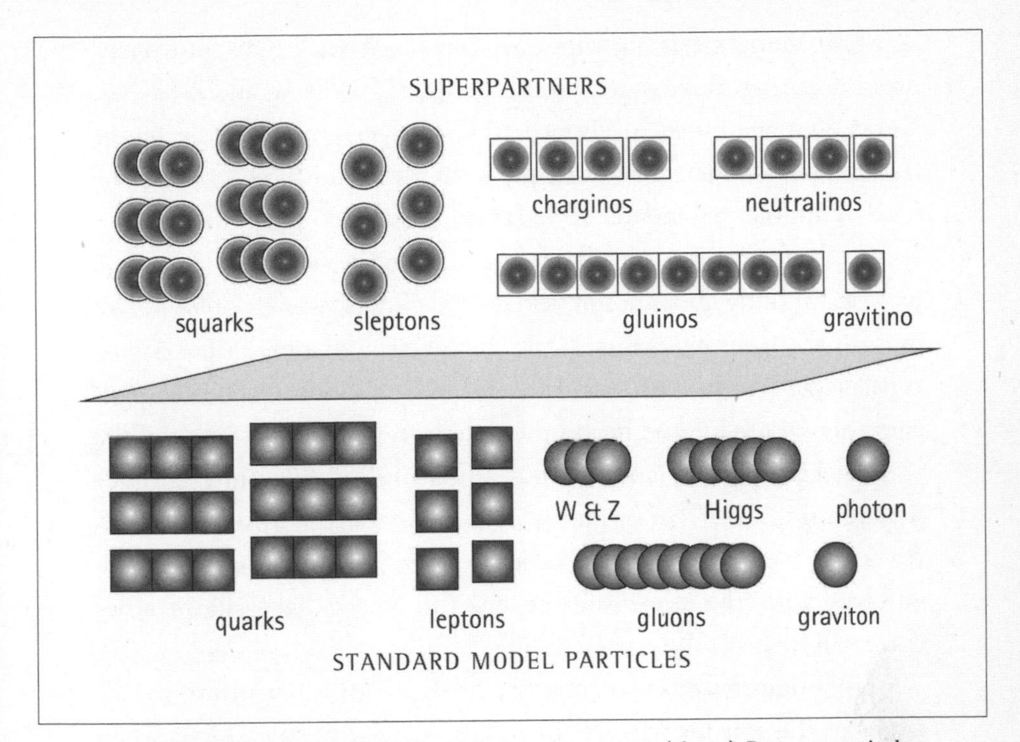

Standard Model particles (below) and their superpartners (above). Bosons are circles, fermions are squares. Three copies of each quark and squark, and eight copies of the gluons and gluinos, represent different colors. The supersymmetric Standard Model has five Higgs bosons rather than the usual one. The superpartners of the W bosons and charged Higgs bosons mix together to make charginos, while the superpartners of the Z, the photon, and the neutral Higgs bosons mix to make neutralinos.

things tractable, people set many of the parameters to zero or at least set them to be equal. As a practical matter, all of this freedom means that it's very hard to make specific statements about what supersymmetry predicts. For any given set of experimental constraints, it's usually possible to find some set of parameters that isn't yet ruled out.

Apart from the search for the Higgs, searching for supersymmetry is probably the highest-priority task of the LHC. Given the messiness of the theory, even if we find it there will be a great challenge in figuring out that supersymmetry is really what we've found. Interestingly, one implication of supersymmetry is that a single Higgs boson is not enough. Remember from Chapter Eleven that the Higgs field in the

Standard Model starts off as four scalar fields of equal mass; after symmetry breaking, three of those fields get eaten by the W and Z bosons, leaving just one Higgs for us to detect. In supersymmetric versions of the Standard Model, however, it turns out for technical reasons that we need to double the amount of scalar fields we start with, from four to eight. (That's not including the fermionic higgsino superpartners; here we're just talking about boson fields.) One of those groups of four gives mass to the up-type quarks, while the other gives mass to the down-type quarks. We still just have three W and Z bosons; when the Higgs gets a nonzero value and breaks the electroweak symmetry, three of the scalar fields are eaten, and that leaves us with five different Higgs bosons running free. That's right: A simple consequence of supersymmetry is that we have five Higgs bosons rather than the usual one. One will have a positive electric charge, one will be negative, and the other three will be neutral.

Five Higgs bosons is obviously a field day for experimenters. This is one of the reasons why the LHC physicists were so cautious when announcing that they had found a new particle at 125 GeV; it could be *a* Higgs boson, without necessarily being *the* Higgs boson. When people try to construct supersymmetric models, it's easy to make one Higgs lighter than all the rest, so maybe we've just discovered that one. However, it's also a generic feature that the lightest Higgs tends to be quite light—usually 115 GeV or less. It's possible to nudge it up to 125 GeV, but it requires a few unnatural-seeming contortions. There is a pressing need for more data, both to get a better handle on the particle that has been discovered, and to keep looking for more.

Having extra particles to detect makes physicists happy, but it doesn't really count as an advantage to supersymmetry as a theory. Here is a more tangible advantage: It helps solve the hierarchy problem.

The hierarchy problem comes about because we expect the effects of virtual particles to push the value of the Higgs field up to the Planck scale. Closer examination, however, reveals that virtual bosons tend to push the Higgs field one way, and virtual fermions push it the other way. In general, there's no reason to expect these effects to cancel each other; usually, subtracting a random big number from another random big number gives a third (positive or negative) big number, not a small one. But with supersymmetry, all that changes. Now there are exactly matching fermion and boson fields, and the effects from their virtual fluctuations can precisely cancel, leaving the hierarchy intact. This is one of the primary motivations physicists have for taking supersymmetry seriously.

Another motivation comes from the idea of WIMP dark matter. In viable supersymmetric models, the lightest superpartner is a completely stable particle with a mass and interaction strength close to the weak scale. If that particle has no electric charge—i.e., if it's a neutralino—it is a perfect candidate for dark matter. A great deal of theoretical work has gone into calculating the relic abundance of neutralinos in different supersymmetric models. Precisely because there are so many new particles and interactions, a wide range of abundances is possible, but it's not hard to get the correct dark-matter density. If superpartners exist at energies accessible to the LHC, we may be able to achieve a spectacular synthesis of particle physics and cosmology. It's good to aim high.

Strings and extra dimensions

String theory is one of the simplest ideas of all time. Just imagine that the elementary constituents of nature, rather than being pointlike particles, are instead small vibrating strings. The concept can be traced back to separate papers in 1968 and 1969 by Yoichiro Nambu, Holger Nielsen, and Leonard Susskind, who independently suggested that certain mathematical relationships in particle scattering could be simply explained if the particles were replaced by strings. As long as the loops or segments of string are sufficiently small, they will look like particles to us. You're not supposed to ask, "What are the strings made of?" just as you were never tempted to ask, "What is an electron made of?" The string-stuff is the fundamental substance out of which other things are made.

The original string theories described only bosons, and they were plagued by an apparently fatal flaw: Empty space was unstable and would quickly dissolve into a cloud of energy. To fix it, string pioneers Pierre Ramond, André Neveu, and John Schwarz showed how to add fermions to the theory. In the process, they ended up inventing one of the first examples of supersymmetry. Thus was "superstring theory" born. To be clear: Viable models of string theory seem to necessarily be supersymmetric, but there are supersymmetric models that aren't necessarily connected to string theory in any way. If we were to find supersymmetric particles at the LHC, it would improve the chance that string theory is on the right track, but it wouldn't be direct evidence for strings.

Superstrings solved the stability problem of the early string models, but they came with a frustrating feature: a massless particle that coupled to the energy of everything. This was annoying because the early goal of string theory was to explain the strong force, and there wasn't any such particle in nuclear interactions. Then in 1974, Joël Scherk and Schwarz pointed out that there is a famous massless particle that couples to the energy of everything: the graviton. Instead of being a theory of the strong interactions, they suggested, maybe string theory is a theory of quantum gravity, as well as all the other known forces—a *theory of everything.*

This idea was originally met with bemused stares, as particle theorists in the 1970s weren't that concerned with gravity. By 1984, however, it was clear that the Standard Model was doing a good job at explaining particle physics, and theorists were looking for new challenges. In that year, Michael Green and Schwarz showed that superstring theory was able to avoid a mathematical consistency challenge that many thought would render the theory nonviable. Just as the electroweak theory burst into popularity once 't Hooft showed it is renormalizable, the string theory bandwagon took off after the Green-Schwarz paper and has been a major part of particle theory in the years since.

There is yet another problem that string theory needs to solve: the dimensionality of spacetime. Quantum field theory is more flexible than string theory, and there are sensible field theories in all sorts of different spacetimes. But superstring theory is more restrictive; early investigations found that the theory naturally wants to live in precisely ten dimensions of spacetime. That's nine dimensions of space and one of time, in contrast with our usual three dimensions of space and one of time. At this point, the faint of heart would be excused for moving on to other ideas.

But string theorists were entranced by the possibility of bringing gravity into the fold of the known forces, and they persevered. They borrowed an old idea that had been investigated in the 1920s by Theodor Kaluza and Oskar Klein: Perhaps some dimensions of space are hidden from our view by being curled up into a tiny ball, too small to be seen or even probed in high-energy particle accelerators. A cylinder like a straw or a rubber hose has two dimensions—up and down the length, and around the circle—but if you look at it from far away it will appear as a one-dimensional line. From that perspective, a faraway cylinder is a line with a tiny compact circle located at each point. Remember that short wavelengths correspond to high energies; if a compact space is sufficiently small, only extremely high-energy particles will even notice it is there.

This idea of "compactification" of extra dimensions became an important part of attempts to connect string theory with observable phenomena. At a fundamental level, there is very little freedom in creating different versions of string theory; work in the 1980s showed that there are really only five string theories. But each of those five features ten dimensions of spacetime, and when we hide six of them we find out that there can be many different ways to perform the compactification. Even though it would take very high energies (presumably of the order of the

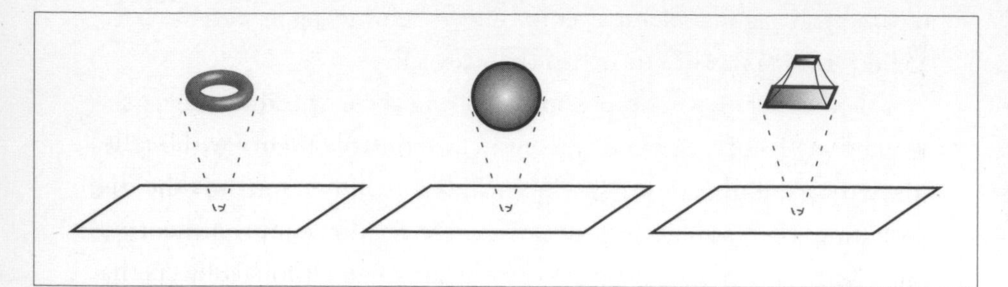

Three different models of compactification. What looks like a point to a macroscopic observer is revealed, on closer inspection, to be a higher-dimensional space. From left to right: a torus (surface of a doughnut), a sphere (surface of a ball), and a warped space stretching between two branes. Realistic compactifications will involve a larger number of extra dimensions, which are hard to illustrate.

Planck energy of quantum gravity, 10^{18} GeV) to directly probe a compact manifold, features of the compactification directly affect the kinds of physics we see at low energies. By "features of the compactification" we mean its volume, its shape, and its topology; compactifying on a torus (the surface of a doughnut) will be very different from compactifying on a sphere (the surface of a ball). And by "the physics we see at low energies" we mean what kind of fermions there are, which forces exist, and the values of the various masses and interaction strengths.

Therefore, while string theory itself is fairly unique, connecting it to experiments has proven to be extremely difficult. Without knowing how the extra dimensions are compactified, it's impossible to say much about what predictions string theory would make for the observable world. This is a pretty general problem with any attempt to apply quantum mechanics to gravity, not just string theory: Direct experimental probes require energies at the Planck scale, and no feasible particle accelerator is going to reach that. That's not to say there can never be data that helps us test models of quantum gravity, but the tests are going to require subtlety rather than brute force.

Branes and the multiverse

In the 1990s, the way people tried to connect string theory with reality underwent a dramatic shift. The impetus for this change was the discovery by Joseph Polchinski that string theory isn't simply a theory of one-dimensional strings. There are also higher-dimensional objects that play a crucial role.

A two-dimensional surface is called a "membrane," but string theorists needed to be able to describe three-dimensional and higher-dimensional objects as well, so they adopted the terminology "2-brane" and "3-brane" and so on. A particle is a zero-brane, and a string is a 1-brane. Using these extra branes, string theorists showed that their theory is even more unique than they thought: All five of the ten-dimensional superstring theories—as well as an eleven-dimensional "supergravity" theory that doesn't have strings at all—are simply different versions of one underlying "M-theory." To this day, nobody really knows what the "M" in "M-theory" is supposed to stand for.

The bad news is that this menagerie of branes nudged string theorists into discovering even more ways to compactify the extra dimensions. Partly this was driven by attempts to find compactifications that featured a positive amount of vacuum energy, which was demanded by the 1998 discovery that the universe is accelerating—one of the rare times that progress in string theory was instigated by experiment. Lisa Randall and Raman Sundrum used brane theory to develop an entirely new kind of compactification, in which space "warped" in between two branes. This led to a rich variety of new approaches to particle physics, including new ways of addressing the hierarchy problem.

It also, unfortunately, seemingly dashed remaining hopes that finding the "right" compactification would somehow allow us to connect string theory with the Standard Model. The number of compactifications we're talking about is hard to estimate, although numbers like 10^{500} have been bandied about. That's a lot of compactifications, especially when the task before you is to search through all of them looking for one that matches the Standard Model.

In response, some proponents of string theory took a different tack: Rather than finding the one true compactification, they imagine that different parts of spacetime feature different compactifications, and that every compactification is realized somewhere. Because compactifications define the particles and forces seen at low energies, this is like having different laws of physics in different regions. We can then call each such region a separate "universe," and the whole collection of them is the "multiverse."

It might seem that such a scheme gives up on any pretense of making testable predictions. It's certainly difficult, but advocates of the multiverse argue that there is still hope. In many parts of the multiverse, they argue, conditions are so inhospitable that no intelligent life can possibly arise. Maybe there are no appropriate forces, or the vacuum energy could be so high that individual atoms would be torn apart by the expansion of the universe. One problem is that we don't have a very good understanding of the conditions under which life can form. If we can overcome such mundane considerations, however, optimists hold out hope that they can make predictions for what typical observers in the multiverse would actually observe. In other words, even if we don't see other "universes" directly, we might be able to use the idea of the multiverse to make testable predictions. The "anthropic principle" is the idea that there is a strong selection effect that limits the conditions we can possibly observe to those that are compatible with our existence.

It's an ambitious plan, and possibly doomed to failure. But people try, and in particular they have applied this idea to properties of the Higgs boson. These are treacherous waters; back in 1990, Mikhail Shaposhnikov and Igor Tkachev tried to predict the value of the Higgs mass under certain anthropic assumptions and came up with the answer 45 GeV. That's clearly incompatible with the data as we now understand it, so something was wrong about those assumptions. Under different assumptions, in 2006, another group predicted a value of 106

GeV; closer, but still no cigar. Now that we have a Higgs boson at 125 GeV, it is unlikely that many predictions will be published that don't somehow manage to reach that value.

To be fair, we need to mention the most impressive success of anthropic reasoning: predicting the value of the vacuum energy. In 1987, more than ten years before the discovery of the acceleration of the universe, Steven Weinberg pointed out that a very high (or large and negative) vacuum energy would inhibit the formation of galaxies. Therefore, most observers in a multiverse should see small but nonzero values of the vacuum energy. (Zero is allowed, but there are more nonzero numbers than numbers equal to zero.) The value we think we have observed is perfectly consistent with Weinberg's prediction. Granted, Weinberg was implicitly imagining a multiverse in which only the value of the vacuum energy changed from place to place; if we let other parameters change, the agreement becomes much less impressive.

Despite the pessimistic, even curmudgeonly tone of this section, I believe the multiverse scenario is actually quite plausible. (In *From Eternity to Here*, I suggested that it might be helpful in explaining the low entropy of the early universe.) If string theory or some other theory of quantum gravity allows for different manifestations of local laws of physics in different regions of spacetime, the multiverse might be real, whether we can observe it or not; I'm always an advocate for taking seriously things that might be real. At the current state of the art, however, we are very far from being able to turn the multiverse into a predictive theory for particle physics. We can't let our personal distaste color our judgment of cosmological scenarios, but neither can we let our enthusiasm get in the way of our critical faculties.

Venturing forth

There is much more to be discovered in the realm of the very small, and there are many aspects to particle physics beyond the Standard Model.

Why is there more matter than antimatter in the universe? Several scenarios for generating such an asymmetry involve the cosmological evolution of the Higgs field, so it's plausible that a better understanding of its properties will lead to new insights on this problem. There are also interesting "technicolor" models, according to which the Higgs is a composite particle like the proton rather than something fundamental. Current versions of technicolor tend to be disfavored by other particle-physics data, but studying the actual Higgs itself might very well lead to surprises.

Discovering the Higgs is not the end of particle physics. The Higgs was the final piece of the Standard Model, but it's also a window onto physics beyond that theory. In the years to come, we'll be using the Higgs to search for (and hopefully study) dark matter, supersymmetry, extra dimensions, and whatever other phenomena prove to be needed to fit the new data that is rapidly coming in. The Higgs discovery is the end of one era and the beginning of another.

THIRTEEN
MAKING IT WORTH DEFENDING

In which we ask ourselves why particle physics is worth pursuing, and wonder what comes next.

Robert Wilson, the physicist who was in charge of building Fermilab, was dragged before the Congressional Joint Committee on Atomic Energy in 1969 to help senators and representatives understand the motivation behind the multimillion-dollar project. It was a turning point in the history of particle physics in the United States; the Manhattan Project had given physicists easy access to influence and funding, but it was unclear how the search for new elementary particles was going to lead to anything as immediately valuable as a new kind of weapon. Senator John Pastore of Rhode Island asked Wilson directly: "Is there anything connected with the hopes of this accelerator that in any way involves the security of the country?"

Wilson answered with equal directness: "No, sir, I don't believe so."

We can imagine that Pastore was a bit taken aback by this answer; presumably he expected to hear a song and dance about how Fermilab played a crucial role in keeping up with the Soviets, the kind of argument that was trotted out for all kinds of purposes in that era. He asked if there was really nothing at all, to which Wilson simply replied, "Nothing at all." But you don't get to be senator without being at least a little stubborn, so Pastore tried a third time, just to ensure he had heard correctly: "It has no value in that respect?"

Wilson was no dummy; he realized he was expected to provide a little bit more if he wanted Congress to fund his ambitious but esoteric endeavor, but he refused to back off from his original point. His answer is one of the most-remembered quotes in the long history of scientists trying to explain why they do what they do:

> It has only to do with the respect with which we regard one another, the dignity of man, our love of culture. It has to do with: Are we good painters, good sculptors, great poets? I mean all the things we really venerate in our country and are patriotic about. It has nothing to do directly with defending our country except to make it worth defending.

Big Science is not cheap. The Large Hadron Collider has cost about $9 billion, almost all of which came ultimately from taxes collected in countries around the world. The people who paid that money have a right to know what they are getting for their investment. It's the duty of the scientific community to be as honest and convincing as possible about the rewards of basic research.

Some of those rewards come in the form of technological breakthroughs. But ultimately, those are not the most important rewards; what matters most is the knowledge that is brought back to us by these extremely ambitious experiments.

Not everyone agrees. Steven Weinberg, who has been a tireless advocate for investment in basic science, recalls a telling anecdote.

> During the debate over the SSC, I was on the Larry King radio show with a congressman who opposed it. He said that he wasn't against spending on science, but that we had to set priorities. I explained that the SSC was going to help us learn the laws of nature, and I asked if that didn't deserve a high priority. I remember every word of his answer. It was "No."

It's not an uncommon attitude. But it's an impoverished perspective, one that misses the bigger picture. Basic science might not lead to immediate improvements in national defense or a cure for cancer, but it enriches our lives by teaching us something about the universe of which we are a part. That should be a very high priority indeed.

When do I get my jet pack?

None of which is to say that we wouldn't *like* to have useful technological applications of the work being done in modern particle physics. Scientists are quick to point out that basic research—scientific investigation carried out purely for its own sake, rather than in pursuit of immediate applications—has very often ended up having enormously practical implications, even if they were unanticipated at the time. From electricity to quantum mechanics, the pages of history are strewn with ideas that once were abstract and impractical, only to later become central to technological progress. As a result, whenever new scientific discoveries are made, people want to know: When do I get my jet pack?

Can we imagine that something similar will be the fate of research at the LHC? As Yogi Berra once said, making predictions is hard, especially about the future. However, we can admit that what we find at the LHC might have a very different character from the fundamental physics of previous centuries. It's possible that any particles we discover at the LHC will literally never be put to good use in practical devices.

That's not just pessimism; it flows from the particular kinds of things we might hope to discover. When Benjamin Franklin was studying electricity or Heinrich Hertz was producing radio waves, they weren't creating things that didn't already exist in the world. Electricity and radio waves are all around us, even if we discount all the artificial sources of them. Scientists in that era were learning to manipulate mysterious features of the readily accessible world, and it's not surprising that the knowledge they discovered later became technologically useful. At the LHC, by contrast, we are literally creating particles that don't exist in our everyday environments. There are good reasons for that. The particles are typically very massive, so it requires an enormous amount of energy to make them. And they are either very weakly interacting, so it's hard to capture or manipulate them (like neutrinos), or they are extremely short-lived, so they decay before they can be put to much good use.

Take the Higgs boson as an example. It's not easy to make a Higgs boson—the only way we know is to have a particle accelerator several miles long. We can certainly imagine technological improvements that would give you a pocket-size device able to reach such high energies; nobody has any idea how to do it, but it doesn't violate the laws of physics. But even if you had a handy iHiggs boson producer, what would it be good for? Every Higgs you make decays in less than a zeptosecond. It's hard to imagine any application of those bosons that wouldn't be carried out more efficiently by some other kind of particle.

This argument isn't airtight, of course. Muons are unstable particles, and they have found potential technological applications, from catalyzing nuclear fusion to searching for hidden chambers in pyramids. But the muon has a lifetime of about one millionth of a second, much longer than a Higgs boson. Neutrinos are stable but weakly interacting, and some farsighted folks have imagined using them for communication purposes. If we were feeling especially expansive, we might imagine discovering dark-matter particles that could find similar uses. It's not a place I would recommend investing a lot of money, however.

Warp drive and levitation

Because the Higgs boson is responsible for giving particles mass, people sometimes wonder whether mastering its properties will allow us to make things lighter or heavier. Or worse. The day after the July 4 announcement of the Higgs discovery, Canada's *National Journal* printed a bold headline: HIGGS BOSON FIND COULD MAKE LIGHT-SPEED TRAVEL POSSIBLE, SCIENTISTS SAY. None of the scientists quoted in the article said anything of the sort, but I suppose it's possible that some scientists somewhere did say that at some point in time.

Using the Higgs to make things lighter or even massless is pretty much a nonstarter, for a few reasons. Most obvious, the large majority of the mass in ordinary objects doesn't come from the Higgs; it comes from the strong-interaction energy inside protons and neutrons. But more important, it's not really the Higgs boson that gives mass to the quarks and charged leptons, it's the Higgs field lurking in empty space. If you wanted, for example, to change the mass of an electron, it's not a matter of shooting Higgs bosons at it; you would have to change the value of the background Higgs field.

That's easier said than done. For one thing, while we can imagine changing the Higgs field, we have no idea how to actually do it. For another, it would require an absurd amount of energy. Let's imagine that we figured out a way to displace the Higgs field from its regular value (246 GeV) all the way to zero, inside some small but macroscopic volume of space. The usual value the Higgs field has is the state of minimum energy it can be in; pushing it back to zero means that our small volume is now packed with energy. From $E = mc^2$, that means it has mass. A quick calculation reveals that a region the size of a golf ball, inside of which the Higgs field is displaced to zero, would have approximately the same mass as the entire earth. If we were to make it much bigger than that, there would be so much mass inside a small space that the whole volume would collapse to make a black hole.

Finally, even if you somehow managed to turn off the Higgs field in, say, your body, it's not just that you would suddenly become lighter. Certain elementary particles would become lighter—the electrons and quarks—and the broken symmetry of the weak interaction would be restored. As a result, the atoms and molecules in your body would fall into completely different configurations, mostly just disintegrating altogether and releasing a huge amount of energy. Decreasing the Higgs field wouldn't put you on a diet; it would make your body explode.

So, don't be looking forward to Higgs-powered levitation devices anytime soon. On the other hand, it remains entirely possible that new discoveries at the LHC will lay the groundwork for future applications in ways we can't currently anticipate. Even if that's not why we pursue them.

Spinoffs

Research in particle physics often does lead to very tangible benefits. Those benefits usually take the form of spinoffs—new technologies that were developed to help meet the challenges posed by the experimental effort itself, rather than direct applications of finding new particles.

The most obvious example is the World Wide Web. Tim Berners-Lee, working at CERN, pioneered the Web when he was trying to develop ways to make it easier for particle physicists to share information. Now it's hard to imagine our world without it. Nobody ever suggested funding CERN because some day they would invent the WWW; it's just a matter of putting smart people into an intense environment with daunting technological challenges, and reaping the benefits from what comes out.

There are many other similar examples. The need for uniquely powerful magnets in particle accelerators has led to noticeable advances in superconducting technology. The ability to manipulate particles has had applications in medicine, food sterilization and testing, and other areas of science such as chemistry and biology. The durable and high-precision detectors that appear in particle experiments have found uses in medicine, radiation testing, and security. The incredible demands on computing power and information transfer between particle physicists have led to advances in computer technology. The list is very long, but the lesson is very clear: Money spent on searching for esoteric particles doesn't just slide down the drain.

It's hard to quantify exactly how efficient it is to invest in fundamental research. Studies by economist Edwin Mansfield suggested that, for society as a whole, it is a wise investment indeed. Mansfield argues that public spending on basic science yields an average return of 28 percent, which almost anyone would be thrilled to get out of their investment portfolio. A number like that is suggestive at best, because the details depend greatly on what industries are studied and what counts as "basic science." But it reinforces the anecdotal impression that, at the cutting edge of science, even the most nonapplied research yields impressive dividends.

The most important spinoff of basic research isn't technological at all—it's the inspiration that science provides for people of all ages. Who knows when a certain child is going to hear a news story about the Higgs boson, become intrigued by science, start studying, and end up as a world-class doctor or engineer? When society puts some small fraction of its wealth into asking and answering big questions, it reminds us all of the curiosity we have about our universe. And that leads to all sorts of good places.

The future of particle physics

Weinberg's ornery congressman aside, most people are willing to admit that learning the laws of nature is a worthwhile project. It's reasonable to ask, however, precisely how much we think it's worth. The fate of the Superconducting Super Collider weighs heavily on anyone who contemplates the future of particle physics. We live in an era in which money is tight, and expensive projects need to justify themselves. The LHC is an amazing accomplishment and will hopefully hum along for many years to come, but at some point we will have learned everything it has to teach us. What then?

The problem is that, while the overwhelming majority of worthwhile scientific projects are much less expensive than high-energy particle accelerators, there are certain questions that can't be addressed without such a machine. The LHC cost roughly $9 billion, and it has given us the Higgs boson and, hopefully, will give us much more in the future. If we were limited to spending only $4.5 billion on that project, we wouldn't have discovered half a Higgs boson or taken twice as long; we simply would have found nothing. Making new particles requires high energies and substantial luminosities, which require a large amount of precision equipment and expertise, which cost money. Hanging over all the jubilation for the wonderful performance of the LHC is the very real possibility that it may be the last high-energy accelerator built in our lifetimes.

There is no shortage of plans for possible next steps forward, if the money can be found. The LHC itself could be upgraded to higher energies, although that seems like a stopgap solution. More attention has been focused on the possibility of a new linear collider (in a straight line, rather than a ring), which would collide electrons and positrons. One proposal has been dubbed the International Linear Collider (ILC), which would be more than twenty miles long and reach energies of either 500 GeV or 1 TeV.

That sounds like less energy than the LHC, which might seem like a step backward, but electron-positron colliders work in a different mode from that of hadron colliders. Rather than throwing as much energy as possible into collisions and seeing what comes out, electron-positron machines are ideal for precision work, which can be achieved by aiming at precisely the energy required to produce a specific new particle. Now that we believe the Higgs is at 125 GeV, it provides a tempting target for physics at a linear collider.

Cost estimates for the ILC range anywhere from $7 billion to $25 billion, and possible sites have been explored in Europe, the United States, and Japan. The project would clearly require major international collaboration and necessitate as much political acumen as experimental physics know-how. An alternative proposal, the Compact Linear Collider (CLIC), has been developed at CERN. It would actually be shorter but reach higher energies through the use of innovative (and therefore riskier) technologies. In 2012, studies for the two competing projects were brought together under a single umbrella. The new leader of the combined effort will be Lyn Evans, who didn't get to enjoy much of a retirement after stepping down from heading the LHC team. It will be Evans's job to decide on the most promising technology for moving forward, as well as to juggle the competing interests of different countries who would love to host a new collider (but don't want to pay for it).

One of the persistent themes you hear when talking to anyone who has been involved with the LHC is the inspirational success of its international collaboration. Scientists and technicians of many different nationalities and ages and backgrounds have come together to build something larger than themselves. If our larger society can summon the willpower to put substantial resources into new facilities, the future of particle physics is bright. But for that to happen, scientists have to convey the interest and importance of what they do. We can't sell particle physics on the basis that it might someday cure Alzheimer's or lead to portable teleportation devices. We have to tell the truth: We want to discover how nature works. How much that's worth is for the human race as a whole to decide.

Wonder

Interviewing my fellow physicists for this book, I was struck by how many were fascinated by the arts before they eventually turned to science. Fabiola Gianotti, Joe Incandela, and Sau Lan Wu all studied art or music when they were young; David Kaplan was a film major.

It's not a coincidence. Even though our quest to understand how nature works often leads to practical applications, that's rarely what gets people interested in the first place. Passion for science derives from an aesthetic sensibility, not a practical one. We discover something new about the world, and that lets us better appreciate its beauty. On the surface, the weak interactions are a mess: The force-carrying bosons have different masses and charges, and different interaction strengths for different particles. Then we dig deeper, and an elegant mechanism emerges: a broken symmetry, hidden from our view by a field pervading space. It's like being able to read poetry in the original language, instead of being stuck with mediocre translations.

I was recently helping out with a TV show that was trying to explain the Higgs boson. When you do TV, words never suffice; you need compelling images. If you're trying to explain subatomic phenomena, the only way to get compelling images is to reach for an analogy. So here's what I came up with: Imagine little robots scooting about on the floor of a vacuum chamber. Each robot is equipped with a sail, but the sails come in all different sizes, from fairly large to quite small. We first film the robots when the chamber has been evacuated; they all move at the same speed, since the sails are completely irrelevant when there's no air for them to feel. But then we let the atmosphere into the chamber and film them moving again. Now the robots with tiny sails still move quickly, while those with large sails are much more sluggish. Hopefully the analogy is clear. The robots are particles, and the sails are their couplings to the Higgs field, which is represented by the air. In a vacuum, where there is no air, the robots are all symmetric and move at the same speed. Filling the chamber with air breaks the symmetry, just like the Higgs field does. You could even draw an analogy between sound waves in the air and the Higgs particle.

Since I'm a theoretically minded person myself, nobody wants to put me in charge of robots, so I consulted with some of my colleagues at Caltech in engineering and aeronautics. When I explained what we wanted to do, the response was universal: "I have no idea what the Higgs boson is, or whether that's a good analogy, but it sounds *awesome*."

At heart, science is the quest for awesome—the literal awe that you feel when you understand something profound for the first time. It's a feeling we are all born with, although it often gets lost as we grow up and more mundane concerns take over our lives. When a big event happens, like the discovery of the Higgs boson at the LHC, that child-like curiosity in all of us comes to the fore once again. It took thousands of people to build the LHC and its experiments and to analyze the data that led to that discovery, but the accomplishment belongs to everyone who is interested in the universe.

Mohammed Yahia writes *Nature* magazine's *House of Wisdom* blog, dedicated to science in the Middle East. After the July 4 seminars announcing the discovery of the Higgs, he celebrated the universality of the scientific impulse.

> As people across the Arab world are all dealing with their politics, revolutions, human rights issues and uprisings, science speaks to all of us equally and we become one. The only two human endeavours that are cross-boundary at this massive scale are art and science.

On July 4, 2012, only hours after the seminars that announced the discovery of the Higgs boson to the waiting world, Lyn Evans was asked what he hoped young people would take from the news. His response was immediate: "Inspiration. These big flagship projects have to be inspirational. When we were young there were lots of things going on—putting a man on the moon. Exciting young people in science is essential." They've succeeded.

Meaning and truth

Particle physics can trace its roots back to the atomists of ancient Greece and Rome. Philosophers such as Leucippus, Democritus, Epicurus, and Lucretius developed an understanding of the natural world based on the idea that matter and energy represented different arrangements of a small number of fundamental atoms. They were not scientists in the modern sense of the word, but some of their insights fit quite well with how we think about the universe today.

The ancient world didn't recognize the strict boundaries we have erected to separate academic disciplines in the contemporary university, so as philosophers they were as interested in ethics and the meaning of life as they were in material reality. As with their understanding of atoms, not all of their conclusions hold up from our perspective today, but many of their ideas still remain relevant. They attempted to follow the logical consequences of their atomic view of the world. If reality is simply the interplay of atoms, where are we to find purpose and meaning? Epicurus, in particular, responded to this challenge by locating value in life as we actually live it here on earth, encouraging his followers to be tranquil in the face of death, to value friendship highly, and to seek pleasure in moderation.

Science is ultimately a descriptive enterprise, not a prescriptive one. It tells us what happens in the world, not what should happen or how to judge what happens. Knowing the mass of the Higgs boson doesn't make us better people, or help us decide which charity to support. Nevertheless, the practice of science has crucial lessons for how we live our lives.

The first lesson is that we are part of the universe. Everything in the human body is successfully described by the Standard Model of particle physics. The heavier elements that are so crucial to our biochemistry were formed by nuclear fusion inside stars, leading to Carl Sagan's dictum, "We are star stuff." Knowing that our atoms obey the Standard Model isn't very helpful when it comes to real-world problems of politics, psychology, economics, or romance; but any ideas you have along those lines must at least conform to what we know about the behavior of elementary particles.

We are part of the universe that has developed a remarkable ability: We can hold an image of the world in our minds. We are matter contemplating itself. How is that possible? Particle physics doesn't give us the answer, but it's a basic ingredient in the larger story in which the answer arises. With the discovery of the Higgs boson, our understanding of the physics underlying everyday reality is complete. There is plenty of room for new particles and forces, but only ones that interact so weakly or briefly with ordinary matter that we can't perceive them without billions of dollars' worth of apparatus. This is a towering achievement in human intellectual history.

The other lesson of science is that nature doesn't let us fool ourselves. Science proceeds by making guesses, which it dignifies by calling them "hypotheses," and then testing those guesses against the data. The process might take decades or longer—and what qualifies as "the best explanation for the data" is a notoriously knotty problem—but ultimately, the experiments have the final say. It doesn't matter how beautiful your idea is, or how many awards you've won, or how many IQ points you have; if your theory contradicts the data, it's wrong.

This is a good-news/bad-news situation. The bad news is that science is hard. Nature is unforgiving, and most theories one can imagine proposing (and indeed, the vast proportion of theories that actually are proposed) will turn out to be incorrect. But the good news is that nature, that strict taskmaster, gradually guides us to ideas we never would have invented through pure thought alone. To paraphrase Sidney Coleman, a thousand philosophers thinking for a thousand years would never have invented anything as strange as quantum mechanics. It's only because the data force us into corners that we are inspired to create the highly counterintuitive structures that form the basis for modern physics.

Imagine that a person in the ancient world was wondering what made the sun shine. It's not really credible to imagine that they would think about it for a while and decide, "I bet most of the sun is made of particles that can bump into one another and stick together, with one of them converting into a different kind of particle by emitting yet a third particle, which would be massless if it wasn't for the existence of a field that fills space and breaks the symmetry that is responsible for the associated force, and that fusion of the original two particles releases energy, which we ultimately see as sunlight." But that's exactly what happens. It took many decades to put this story together, and it never would have happened if our hands weren't forced by demands of observation and experiment at every step.

The flip side is that once the data put us on the right track, science is capable of extraordinary leaps into the future. In the 1960s, physicists constructed a unified theory of the electromagnetic and weak interactions, based on some general principles that previous experiments had validated and some specific observational facts, such as the absence of massless weak-force-carrying bosons. That theory made a prediction: There should be a new massive particle, the Higgs boson, that couples to the known particles in certain definite ways. In 2012, a full forty-five years after Steven Weinberg's 1967 paper put the theory together, that prediction came true. The human intellect, guided by nature's clues, was able to figure out a deep truth about how the universe works. And on the basis of that insight we hope to see even further in the years to come.

When I talked with JoAnne Hewett about what makes a successful physicist, one word kept recurring: "persistence." Individual scientists require persistence to stick with tough problems, and society as a whole needs to be willing to support costly long-term projects to tackle our hardest questions. When it comes to understanding the architecture of reality, the low-hanging fruit has been picked. The easy part is over.

The questions we are faced with are difficult ones, but if recent history is any guide, a combination of dogged effort and occasional flashes of insight should be able to get us there. The construction of the Standard Model may be complete, but the task of incorporating the rest of reality into human understanding remains before us. If it weren't a challenge, it wouldn't be so much fun.

APPENDIX ONE
MASS AND SPIN

The first thing we say about the Higgs field is that it gives mass to other particles. In this appendix, we're going to explain what that means, somewhat more carefully than we do in the main text. Nothing here is absolutely necessary, but it may clarify a thing or two.

So: Why do we *need* a field whose job it is to give mass to other particles? Why can't the particles just have mass without any help?

We can certainly imagine massive particles without the Higgs field being involved at all. But the particles of the Standard Model are of a special type that doesn't allow for that to happen. There are two different sets of particles that get mass from the Higgs field: the force-carrying W and Z bosons of the weak interactions, and the electrically charged fermions (electron, muon, tau, and all the quarks). The way the bosons get masses is different in detail from the way the fermions get masses, but the underlying motto is the same in both cases: There is a symmetry that seems to prohibit any mass at all, and the Higgs field breaks that symmetry. To understand how that happens, we need to talk about the *spin* of elementary particles, which we've been skirting around thus far in the book.

Spin is one of the fundamental defining features of a particle in quantum mechanics. The phrase "quantum mechanics" itself, although not the most accurate terminology ever invented, stems from the fact that certain things come only in discrete packets, not in any possible amount. For example, the energy of an electron that is bound to an atomic nucleus is only allowed to take on certain specific values. The same is true for a quantity known as the "angular momentum," which is a way of characterizing how fast one object is rotating or moving around another object. The rules of quantum mechanics tell us that angular momentum is quantized: It can come only in fixed multiples of a fundamental value. The minimum unit of angular momentum is given by Planck's constant *h*, a fundamental quantity of nature, divided by two pi. This quantity is so important that it gets its own funky orthography, and is called $\hbar$, pronounced "*h*-bar." Planck invented his original constant *h* back in the earliest days of quantum mechanics, but it turns out that $\hbar$ is much more useful, so we often simply call that "Planck's constant." Numerically, $\hbar$ is equal to about 6.58×10^{-16}, in units of electron volts times seconds.

Imagine you have a spinning top that you can manipulate very precisely. You make it spin more and more slowly, and observe what it's doing as accurately as you please. What you will find is that as you slow the top down, only discrete rotational speeds are allowed; the rotation of the top will suddenly change from one speed to another, like the second hand on a quartz watch suddenly hopping from one second to the next. Eventually you will hit a slowest-possible rotation, when the total angular momentum of the top is equal to $\hbar$. The reason why you don't notice this when you are watching Olympic skaters spinning on the ice is that the minimum rotation is very slow: A toy top with angular momentum $\hbar$ would take a hundred trillion times the age of the universe to complete a full turn.

The spin of a top has angular momentum because the atoms in the top are literally rotating around some central axis. One of the consequences of quantum mechanics is that individual particles can also have "spin," even though they're not really rotating around anything.

The way we know that is because total angular momentum stays constant through time, and we see processes where orbiting particles interact and get turned into particles that aren't orbiting at all. In this case we can conclude that the angular momentum must have gone into the spin of the particle. When we say "spin," we'll always mean this intrinsic quantum-mechanical spin of elementary particles, and when we say "angular momentum," we'll be thinking of the classical phenomenon of one object moving around another one (also known as "orbital" angular momentum).

How spin works

There are a few crucial facts we need to know about particle spin. Every kind of particle has a fixed amount of spin once and for all; they never start spinning faster or slower. Measured in units of $\hbar$, every photon in the universe has a spin equal to one, and every Higgs boson has a spin equal to zero. Spin is an intrinsic feature of the particle, not something that changes as the particle evolves (unless it transforms into another kind of particle).

Unlike regular orbital angular momentum, the smallest unit of spin is one-half $\hbar$, rather than $\hbar$ itself. An electron has a spin of one-half, as does an up quark. Why this is possible is an amusing quirk of quantum field theory, but delving into it would take us even further afield than the rest of this relatively technical appendix.

There is a simple correlation between the spin of a particle and its nature as a boson or fermion. Every boson has a spin that is an integer: 0, 1, 2, etc. (in units of $\hbar$, which we will assume henceforth). Every fermion has a spin that is an integer plus a half: 1/2, 3/2, 5/2, etc. This connection is so close that people often define bosons as "particles with integer spin" and fermions as "particles with half-integer spin." That's not really right; the definition we gave, that bosons can pile on top of one another while fermions take up space, is the true distinction between these two classes of particles. A famous theorem in physics, the "spin-statistics theorem," tells us that particles that can pile on top of one another must have integer spins, and particles that take up space have half-integer spins. At least in a four-dimensional spacetime—but that's all we care about here.

The particles of the Standard Model have very specific spins. All of the known elementary fermions—quarks, charged leptons, and neutrinos—are spin-1/2. The gravitino, hypothetical supersymmetric partner of the graviton, would be spin-3/2, but no gravitinos have ever been observed. The graviton itself is spin-2, uniquely among the elementary particles. The other gauge bosons—the photon, gluons, and the Ws and Z—are all spin-1. (The difference between the graviton and the other force-carrying bosons is ultimately traced to the fact that the symmetry underlying gravity is a symmetry of spacetime itself, while the other forces propagate on top of spacetime.) The Higgs boson, standing apart from all the rest, is spin-0. Particles with zero spin are called "scalars," and the fields from which they arise are called "scalar fields."

It's important to distinguish between the "spin of a particle" and the "value of the spin we measure with respect to some axis." Suppose the angular momentum of the earth spinning on its axis, pointing from the South Pole to the North Pole, is some (large) number. Then we say that we could imagine measuring the angular momentum with respect to an axis pointing in the opposite direction, from north to south. This answer would be minus the original right-side-up answer. The angular momentum hasn't changed, we've just measured it with respect to a different axis. If we look at the original axis from above, a positive spin means we see the object rotating counterclockwise, while a negative spin means it's rotating clockwise. The earth spins counterclockwise from the perspective of someone looking down from the North Pole, so it has a positive spin. (This is known as the "right-hand rule"—if you curl the fingers of your right hand in the direction something is spinning, your thumb points to the axis along which that spin is positive.)

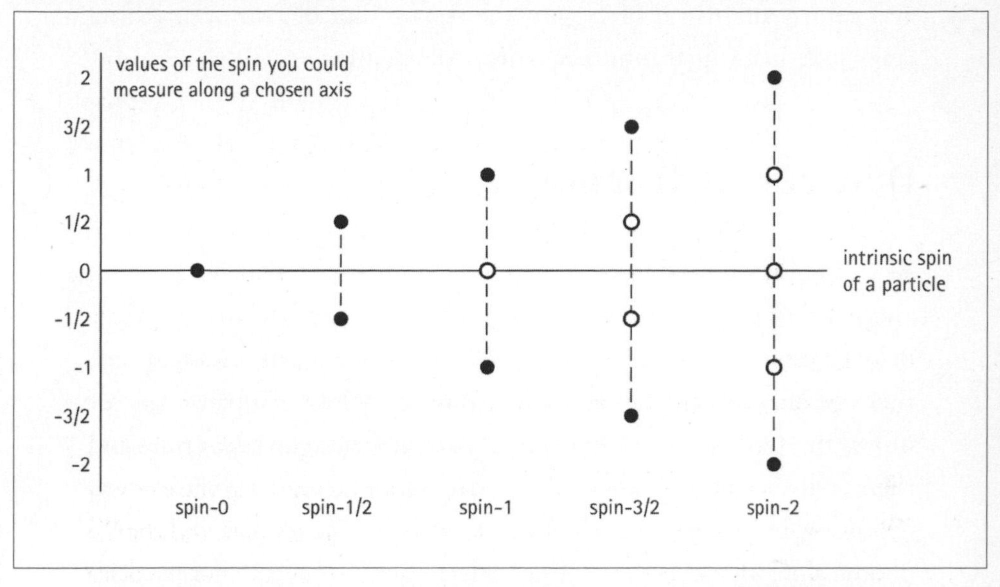

Allowed outcomes of measuring the intrinsic spin of a particle with respect to some axis. Massless particles give only the answers corresponding to filled circles, while massive particles give answers corresponding to both filled and open circles.

We can even consider measuring the angular momentum with respect to a completely perpendicular axis—say, one pointing from one side of the equator to the other. With respect to that direction, the earth isn't "spinning" at all—the North and South Poles stay in the same position with respect to an imaginary axis passing through the equator. So we would say that the spin measured with respect to that axis is zero.

Just as the total spin of a particle is quantized to be some integer or half-integer multiple of $\hbar$, the spin you can measure is also quantized. It must either be equal to the total spin, or minus the total spin, or some number in between separated by integers. For a spin-0 particle, the only possible answer we can get while measuring the spin is 0. For a spin-1/2 particle, we could get +1/2 or −1/2, but that's it; we can't fit any other values in between that are separated by at least one. For a spin-1 particle, we could measure the spin to be +1, −1, or 0. If we measure 0, that doesn't mean the particle isn't spinning; it just means its axis is perpendicular to the one along which we are measuring. No measurement will ever return an answer of 7/13 or the square root of two or anything crazy like that—quantum mechanics doesn't allow it.

Degrees of freedom

At this point we need to draw a distinction between massive particles and massless ones. (See how this is going to connect back to the Higgs field?) It turns out that when you measure the spin of a massless particle, there are only two answers you can possibly get: plus the intrinsic spin or minus the intrinsic spin. (For spin-0 particles those are the same, and there's only one possible answer.) In other words, no matter what axis you choose, when you measure the spin of a massless spin-1 particle like the photon along that axis, you will get either +1 or –1, never 0. For particles with spin-0 or spin-1/2, that doesn't matter; there aren't any missing values. But for higher-spin particles, it matters a lot. When we measure the spin of a photon or graviton, there are only two possible values we can get, but when we measure the spin of a W or Z boson, there are three different values, since it's possible to get 0. In the figure, filled circles represent the results we can get when measuring the spin of a massless particle, while a massive particle could give us any of the filled or open circles.

The reason why this fact is so important is that each of the allowed spin measurements represents a new "degree of freedom." That's physics-speak for "something that can happen independently of other things happening." Since what we're really talking about here are quantum fields, every degree of freedom is a specific way the field can vibrate. For a spin-0 field like the Higgs, there's only one way it can vibrate. For a spin-1/2 field like the electron, it can have two kinds of vibrations, consisting of clockwise or counterclockwise spinning around whatever axis you choose. (Hard to visualize, admittedly.) A massless spin-1 particle like the photon also has just two kinds of vibrations. But a massive spin-1 particle like the Z boson has three kinds of vibrations: With respect to some axis, it could be spinning clockwise, counterclockwise, or not at all.

This might sound like a confusing mess, but if you go back to the discussion of the Higgs mechanism in Chapter Eleven, it helps make sense of what happens when we spontaneously break a local symmetry. Remember that in the Standard Model, we start (before symmetry breaking) with three massless gauge bosons and four scalar Higgs bosons. Count the number of degrees of freedom: two each for the three massless gauge bosons, one each for the scalars, giving $2 \times 3 + 4 = 10$. After symmetry breaking, three of the scalars get eaten by the gauge bosons, which become massive, leaving behind one massive scalar that we observe as the physical Higgs boson. Now count the degrees of freedom again: three each for three massive gauge bosons, plus one for the remaining scalar, giving $3 \times 3 + 1 = 10$. They match. Spontaneous symmetry breaking doesn't create or destroy degrees of freedom, it just jumbles them up.

Counting degrees of freedom helps explain why gauge bosons are massless without the Higgs. The reason gauge bosons exist in the first place is that there is a local symmetry—one that operates separately at every point in space—and we need to define connection fields that relate the symmetry operations at different points. It turns out that you need precisely two degrees of freedom to define this kind of field. (Trust me here. It's hard to think of a sensible explanation that doesn't amount to going through all the math.) When you have a spin-1 or spin-2 particle with just two degrees of freedom, that particle is necessarily massless. The Higgs field is a completely separate degree of freedom; when it gets eaten by the gauge bosons, they now become massive. If there were no extra degrees of freedom lying around, the gauge bosons would have had to remain massless, as they do for the other known forces.

Hopefully this helps explain why physicists were so confident that something like the Higgs must exist, even before it had been discovered. In some sense, it *had* been discovered—three of the four scalar bosons were already there, as the zero-spin parts of the massive W and Z bosons. All we needed to do was find the fourth.

Why fermions are massless without the Higgs

Here is why the fact that fermions have mass is something that demands an explanation in the first place. Notice that the degrees-of-freedom argument we used for the gauge bosons isn't relevant in this case; a spin-1/2 fermion has two possible spin values whether it's massive or massless.

Start by thinking of a massive spin-1/2 particle like the electron. Imagine that it's moving directly away from us, and we measure its spin to be +1/2 along an axis pointing in the direction of its motion. But we can imagine accelerating our own velocity so that we start catching up to the electron—now we're moving toward it. Nothing intrinsic to the electron has changed, including its spin, but its velocity with respect to us has. We define a quantity called the "helicity" of a particle, which is the spin as measured along the axis defined by its motion. The helicity of the electron goes from being +1/2 to −1/2, and all we did is change our own motion—we didn't touch the electron at all. Clearly the helicity isn't an intrinsic feature of the particle; it depends on how we look at it.

Now consider a massless spin-1/2 fermion (like the electron would be without spontaneous symmetry breaking). Let it be moving away from us, and we measure its spin to be +1/2 along an axis defined by its direction of motion, so its helicity is +1/2. In this case, the fermion is necessarily moving at the speed of light (because that's what massless particles always do). Therefore we can't catch up to it and change its apparent direction of motion just by accelerating ourselves. Every observer in the universe will observe this massless particle to have a unique value for its helicity. For massless particles, in other words, the helicity is a well-defined quantity no matter who is measuring it, unlike the case for massive particles. A particle with positive helicity is "right-handed" (spinning counterclockwise as it comes toward us), while a negative-helicity particle is "left-handed" (spinning clockwise as it comes toward us).

And the reason why all this matters is because *the weak interactions couple to fermions of one helicity but not the other.* In particular, before the Higgs comes along to break the symmetry, the massless gauge bosons of the weak interactions couple to left-handed fermions and not right-handed ones, and they also couple to right-handed antifermions and not left-handed ones. Don't ask why that's the way nature works, except that it's what we need to fit the data. The strong force, gravity, and electromagnetism all couple equally well to left- and right-handed particles; but the weak force couples to one but not the other. That also explains why the weak interactions violate parity: Looking at the world through a mirror switches right with left.

Having a force couple to one helicity but not the other clearly doesn't make sense if the helicity is different to observers moving at different speeds. Either the weak force couples to a certain particle, or it doesn't. If the weak force couples only to left-handed particles and right-handed antiparticles, it must be true that such particles have one helicity or the other once and for all. And that can happen only if they move at the speed of light. Which implies, at last, that they must have zero mass.

If you can swallow that, it helps make sense of some of the dancing around we did while first defining the Standard Model. We said that the known fermions come in pairs, which would be symmetric if it weren't for the Higgs lurking in empty space. Up and down quarks form a pair, electrons and electron neutrinos form a pair, and so on. But really it's only the left-handed up and down quarks that form a symmetric pair; there is no local symmetry connecting right-handed up quarks to right-handed down quarks, and likewise for the electron and its neutrino. (In the original version of the Standard Model, neutrinos were thought to be massless, and right-handed neutrinos didn't even exist. Now we know that neutrinos have a small mass, but the status of right-handed neutrinos remains murky.) Once the Higgs fills space, the weak symmetry is broken, and the observed quarks and charged leptons are all massive, with both right- and left-handed helicities allowed.

Now we see why the Higgs is needed in order for Standard Model fermions to have mass. If the weak interaction symmetry were unbroken, helicity would be a fixed property of each fermion, which means that they all would be massless particles moving at the speed of light. It's all because the weak interactions can tell left from right. If that weren't true, there would be no obstacle to fermions simply having mass, with or without the Higgs. Indeed, the Higgs itself is a scalar field with mass, but it's not as if the Higgs gives mass to itself; it simply has mass, since there's no reason for it not to.

APPENDIX TWO
STANDARD MODEL PARTICLES

Throughout the book we've talked about the various particles of the Standard Model, but not always in a systematic way. Here we provide a summary of the particles and their properties.

There are two types of elementary particles: fermions and bosons. Fermions take up space; that is, you cannot put two identical fermions right on top of each other in precisely the same configuration. They therefore serve as the basis for solid objects, from neutron stars to tables. Bosons can be piled on top of one another as much as you like. They therefore are able to create macroscopic force fields, such as the electromagnetic field and the gravitational field.

The fermions

Let's consider the fermions first. There are twelve fermions in the Standard Model that fall into strict patterns. Fermions that feel the strong nuclear force are quarks, while those that don't are leptons. There are six types each of quarks and leptons, arranged into three pairs, each pair forming a generation. It's a rule that the spin of a fermion must

THE PARTICLE ZOO: FERMIONS

Quarks (Strongly interacting; confined in hadrons)		Leptons (Non-strongly interacting; not confined)	
Up-type quarks (charge +2/3)	*Down-type quarks* (charge +1/3)	*Charged leptons* (charge -1)	*Neutrinos* (charge 0)
Top (172 GeV)	Bottom (~4 GeV)	Tau (1.78 GeV)	Tau neutrino (small)
Top (~1 GeV)	Bottom (~0.1 GeV)	Tau (0.106 GeV)	Tau neutrino (small)
Top (~0.002 GeV)	Bottom (~0.005 GeV)	Tau (0.0005 GeV)	Tau neutrino (small)

The elementary fermions, with their electric charges and approximate masses. The masses of neutrinos haven't yet been accurately measured, but they are all lighter than the electron. Quark masses are also approximate; they are hard to measure because quarks are confined inside hadrons.

equal an integer plus one half; all the known elementary fermions are spin-1/2 particles.

There are three up-type quarks, with an electrical charge of +2/3 each. In order of increasing mass, they are the up quark, the charm quark, and the top quark. There are also three down-type quarks, with charge –1/3 each: the down quark, the strange quark, and the bottom quark.

Each type of quark comes in three colors. It would be perfectly legitimate to count each color as a separate kind of particle (in which case there would be eighteen types of quarks, not just six), but because the colors are all related by the unbroken symmetry of the strong interactions, we usually don't bother. All particles with color are confined into colorless combinations known as "hadrons." There are two simple types of hadrons: mesons, which consist of a quark and an antiquark, and baryons, consisting of three quarks, one each of the three colors red, green, and blue. Protons (two ups and one down) and neutrons (two downs and one up) are both baryons. An example of a meson is the pion, which comes in three types: one with positive charge (up plus antidown), one with negative charge (down plus antiup), and one that is neutral (a mixture of up-antiup and down-antidown).

Unlike quarks, leptons are not confined; each one can move by itself through space. The six leptons also come in three generations, each with one neutral particle and one with charge –1. The charged leptons are the electron, muon, and tau. The neutral leptons are the neutrinos: the electron neutrino, the muon neutrino, and the tau neutrino. Neutrino masses are not well understood and don't arise in the same way as those of other Standard Model fermions, so we have essentially ignored them in this book. They're known to be small (less than one electron volt) but not zero.

The twelve different fermions should really be thought of as six different matched pairs of particles. Each charged lepton comes in a pair with its associated neutrino, while the up and down quarks form a pair, as do the charm and strange quarks, and the top and bottom. As an example of this pairing in action, when a W^- boson decays into an electron and an antineutrino, it's always an electron antineutrino. Likewise, when a W^- decays into a muon, it's always accompanied by a muon antineutrino, and so on. (You would like to say the same thing about the quarks, but they actually mix together in subtle ways.) The particles within each pair would actually have identical properties, if it weren't for one sneaky influence lurking in the background: the Higgs field. In the world we see, the particles within each pair have different masses and different electrical charges, but that's only because the Higgs is hiding their underlying symmetric nature.

Is it possible that the quarks and leptons aren't really elementary, and that they are actually made of an even smaller level of particles? Sure, it's possible. Physicists don't have a vested interest in the current particles being truly elementary; they would love to find yet more mysteries hidden within them, and they have spent a great deal of time inventing models along those lines and testing them experimentally.

The hypothetical particles that could make up quarks and leptons even have a name: "preons." What they don't have is any experimental evidence, or, for that matter, any compelling theory. The consensus these days is that quarks and leptons seem to be truly elementary, rather than being composites of some other kind of particle, but we can always be surprised.

The bosons

Now we turn to the bosons, which always have integer spins. The Standard Model includes four types of gauge bosons, each arising from local symmetries of nature, and corresponding to a certain force.

Photons, which carry the electromagnetic force, are massless, neutral, spin-1 particles. Gluons, which carry the strong nuclear force, are also massless, neutral, and spin-1. A major difference is that gluons do carry color, so they are confined inside hadrons just like quarks are. Because of these colors there are actually eight different kinds of gluons,

THE PARTICLE ZOO: BOSONS

	Name	Mass (GeV)	Charge	Spin
Electromagnetism	photon	0	0	1
Strong nuclear force	gluons (8)	0	0	1
Weak nuclear force	W^+, W^-	80.4	+1, -1	1
	Z	91.2	0	1
Gravity	graviton	0	0	2
Higgs	Higgs boson	125	0	0

Force-carrying particles, the bosons. Masses are measured in giga electron volts (GeV).

but once again they are related by an unbroken symmetry, so we don't even bother to give them specific labels.

Gravitons, which carry gravity, are also massless and neutral but have spin-2. Gravitons do interact with gravity themselves—because everything interacts with gravity—but for the most part gravity is so weak that you wouldn't notice. (Things change when you collect a large amount of mass to create a strong gravitational field, of course.) Indeed, the weakness of gravity means that the graviton is mostly irrelevant for particle physics, at least within the Standard Model. Because we don't have a full theory of quantum gravity, and because individual gravitons are almost impossible to detect, people often don't include the graviton as a particle, but there's every reason to believe that it's real.

The weak force is carried by the charged W bosons and the neutral Z bosons. All three are spin-1 but massive, and they decay quickly when they are produced. It's the broken symmetry due to the Higgs field that is responsible for the weak bosons becoming massive and differentiating from one another; if it weren't for the Higgs, the W and Z bosons would be more like gluons, but with only three varieties instead of eight.

Unlike the three forces previously mentioned, the weak force is so feeble that it isn't able to hold any two particles together all by itself. When other particles interact via the weak force, there are essentially only two ways to do it: Two particles can scatter off each other by exchanging a W or Z, or one massive fermion can decay into a lighter fermion by emitting a W, which then decays into other particles itself. Those processes play a crucial role when it comes to looking for new particles at the LHC.

The Higgs itself is a scalar boson, which is to say that it is spin-0. Unlike the gauge bosons, it doesn't arise from a symmetry, and there's no reason to expect that its mass should be zero (or even small). We can talk about a Higgs "force," and it might even be relevant to detecting dark matter in deep underground experiments. But the major interest

WHICH PARTICLES FEEL WHICH FORCES?

Particle	Force				
	Electromagnetism	Strong	Weak	Gravitation	Higgs
quarks	X	X	X	X	X
charged leptons	X		X	X	X
neutrinos			X	X	?
photons				X	
gluons		X		X	
W+, W-	X		X	X	X
Z			X	X	X
graviton				X	
Higgs			X	X	X

Table summarizing which particles (bosons and fermions) interact with which forces. Photons carry the electromagnetic force, but they don't interact directly with themselves, since they are electrically neutral. The origin of neutrino mass is still mysterious, so their interaction with the Higgs is unknown.

in the Higgs comes from the fact that the field on which it is based is nonzero in empty space, and its presence influences other particles by giving them mass.

If you've read this far, you're probably pretty familiar with the Higgs boson.

APPENDIX THREE
PARTICLES AND THEIR INTERACTIONS

This appendix, in which we talk about Feynman diagrams, is also more technical than the main body of the book. Feel free to skip it, or to just look at the pictures. Richard Feynman himself, when he first invented the diagrams, thought it would be hilarious if someday these little scribbles were all over the place in the physics research journals. That hilarity has come to pass.

Feynman diagrams are a simple way to figure out what can happen when elementary particles get together to interact. Let's say you want to ask whether a Higgs boson can decay into two photons. You know that photons are massless, and that the Higgs interacts only with particles that have mass, so your first guess might be that it doesn't happen. But by concatenating Feynman diagrams, you can discover processes in which virtual particles can connect the Higgs to photons. A physicist will then go further, using those diagrams to calculate the actual probability that such an event will occur; each diagram is associated with a specific number, and we add up all the different diagrams to get the final answer. We're not playing the role of professional physicists, but it's still helpful to see the various allowed interactions portrayed in Feynman-diagram language. There are many rules that go along with these diagrams; we'll delve into them just enough to get an idea of what's going on, but if you want to be precise, it will behoove you to consult a textbook on particle physics or quantum field theory.

Some basic principles: Each diagram is a cartoon of particles interacting with one another and changing identity, with time running from left to right. The incoming particles at the far left of a diagram, and the outgoing particles at the far right, are "real"—they have the mass that we've listed in the Particle Zoo tables in Appendix Two. Particles that exist only inside a diagram, not sneaking out to either side, are "virtual"—their mass can be anything at all. That's worth emphasizing: Virtual particles aren't real particles, they're just bookkeeping devices that indicate how quantum fields are vibrating in the course of a particle interaction.

We'll portray fermions with solid lines, gauge bosons with wavy lines, and scalar bosons (such as the Higgs) with dashed lines. Fermion lines never end—they either travel in closed loops, or they stretch to the beginning and/or end of the diagram. Boson lines, in contrast, can easily come to an end, either on fermion lines or on other boson lines. A place where lines come together is called a "vertex." At each vertex, electric charge is conserved; so if an electron emits a W boson to turn into a neutrino, we know it must be a W^-. The total number of quarks and the total number of leptons (where antiparticles count as –1) are also conserved at each vertex. We can take any line and turn it backward if we exchange particles with antiparticles. So if an up quark can convert to a down quark by emitting a W^+, an antidown can convert to an antiup by the same means.

We'll start by writing down the basic diagrams of the Standard Model. More complicated diagrams can be constructed by combining these fundamental building blocks in various ways. We're not going to be completely comprehensive, but hopefully enough that the basic pattern becomes clear.

First, consider what can happen to a single fermion coming in from the left. Fermion lines can't end, so some sort of fermion has to come out the other side. But we can spit out a boson. Essentially, if a fermion feels a certain force, it can emit the boson that carries that force. Here are some examples.

Every particle feels gravity, so every particle can emit a graviton. (Or absorb a graviton, if we run the diagram backward; like photons and the Higgs, gravitons are their own antiparticles.) Even though we're drawing a straight line as if the particle is a fermion, there are equivalent diagrams for all the bosons as well.

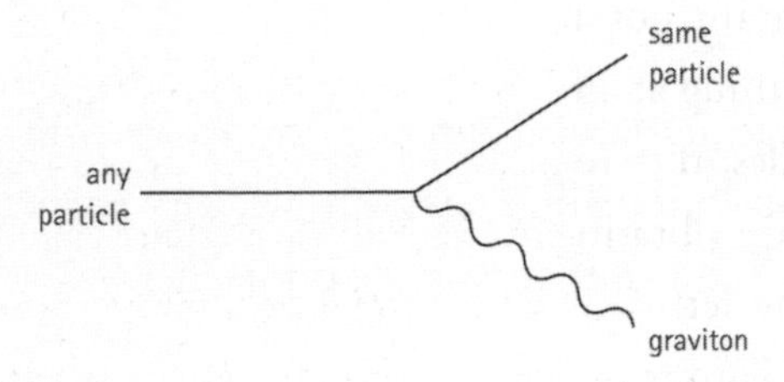

Notice that this diagram, and several of the ones to follow, describe a particle emitting another particle while remaining itself unchanged. That can never happen all by itself, because it wouldn't conserve energy. All diagrams of this sort must be embedded as part of some bigger diagram.

Electromagnetism, unlike gravity, is felt only directly by charged particles. An electron can emit a photon, but a neutrino or a Higgs cannot. They can do so indirectly, through more complicated diagrams, but there's no simple vertex that does the trick.

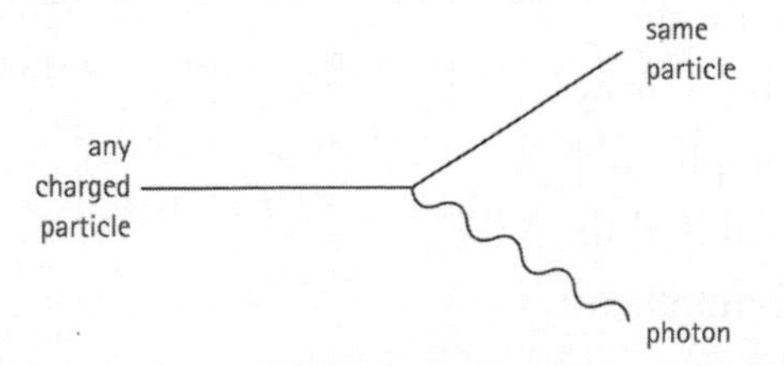

Likewise, any strongly interacting particles (quarks and gluons) can emit gluons. Note that gluons are strongly interacting, while photons are not electrically charged—there is a three-gluon vertex, but no three-photon vertex.

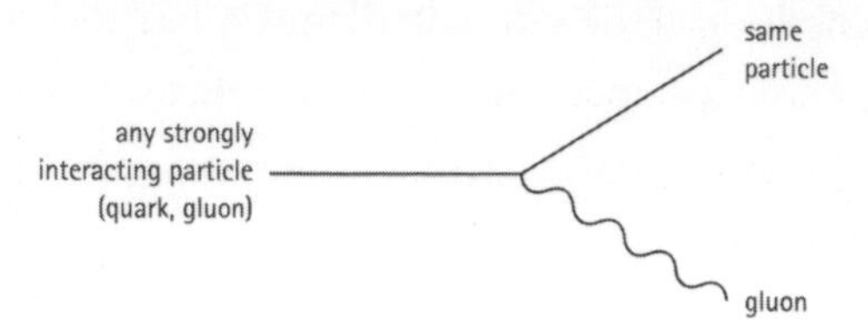

Now we come to the weak interactions, where things are a bit messier. The Z boson is actually pretty simple; any particle that feels the weak interactions can emit one and go on its merry way. (Again, as part of a bigger diagram.)

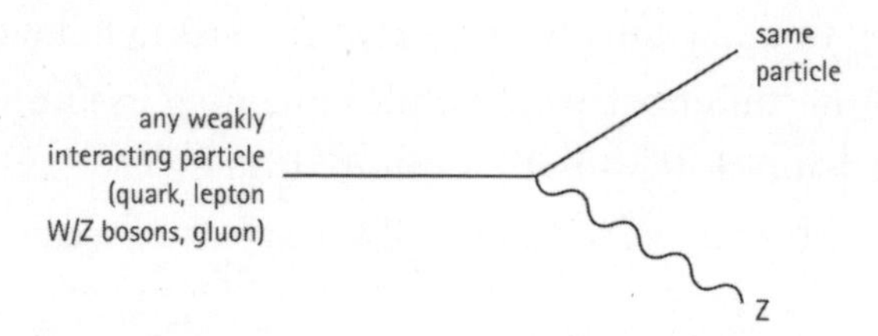

Once we get to the W bosons, things are a bit more complicated. Unlike the other bosons we've just considered, the Ws are electrically charged. That means they can't be emitted without changing the identity of the particle emitting them; if they did, charge wouldn't be conserved. So the W bosons serve to convert between up-type quarks (up, charm, top) and down-type quarks (down, strange, bottom), as well as between the charged leptons (electron, muon, tau) and their associated neutrinos.

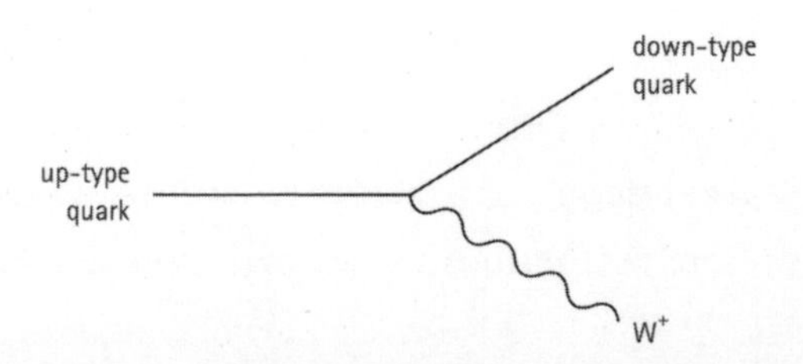

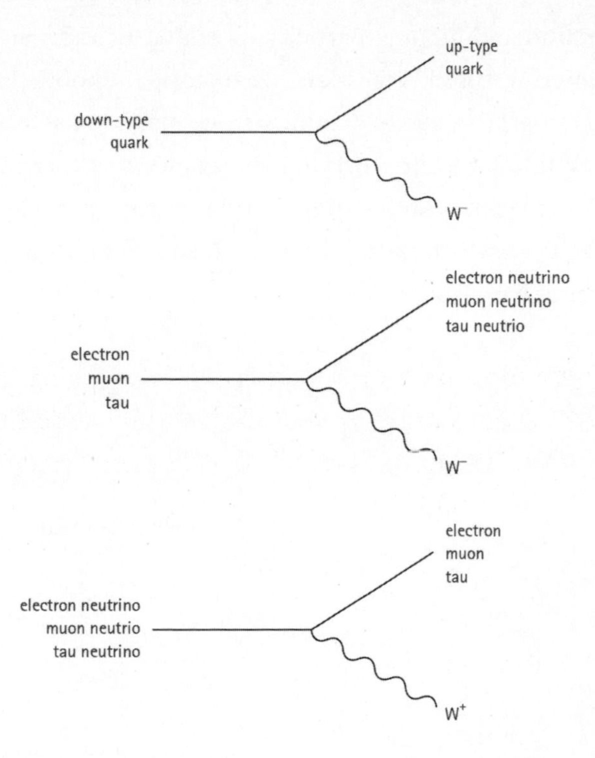

The Higgs boson is much like the Z: Any particle that feels the weak interaction can emit one.

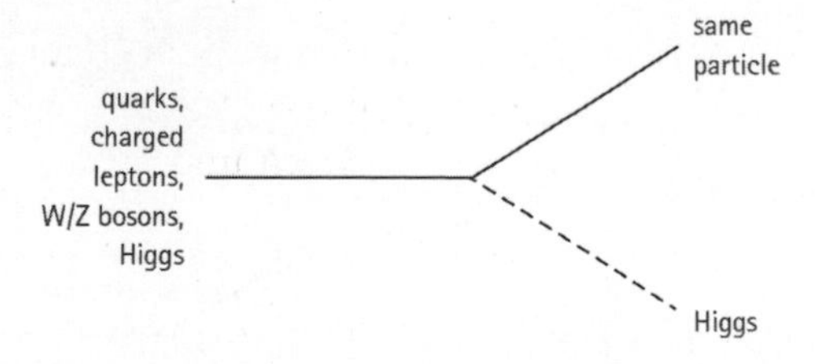

Now we turn to bosons coming in. They can emit another boson, or they can split into two fermions. However, since a fermion line can never end, a boson has to split into one fermion and one antifermion; that way the total number of fermions at the end is zero, just like it was at the beginning. Here we have a multitude of examples. Note that these are all related to diagrams we've already drawn, just by moving lines around and flipping particles to antiparticles where appropriate. If the entering boson is massless, we once again know that it can only be used as part of a bigger diagram, since massless particles can't decay into massive ones while satisfying conservation of energy. (One way to see that is that the combination of two massive particles must have a "rest frame" in which the total momentum is zero, while a single massless particle has no state of rest.)

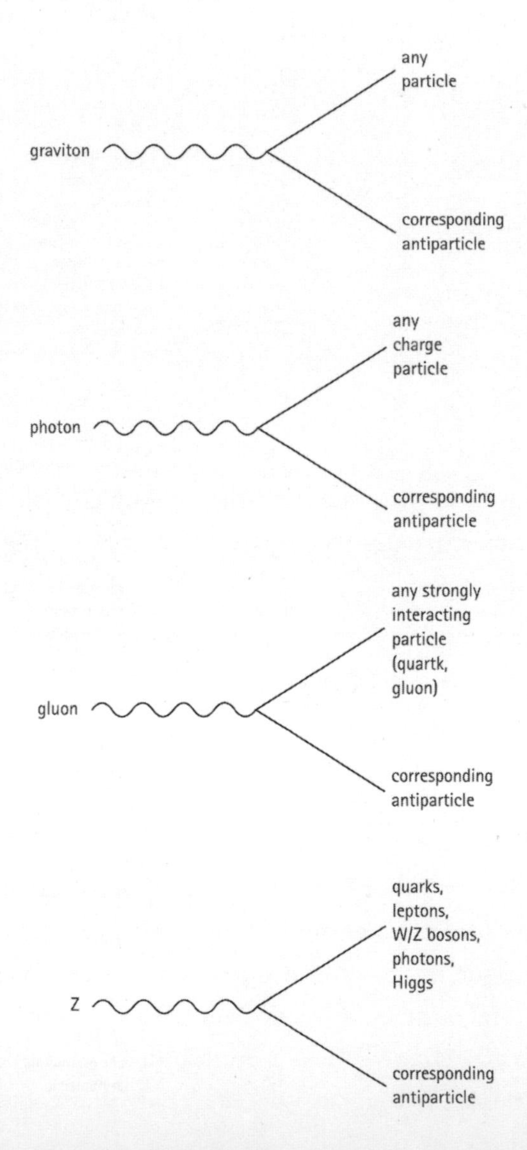

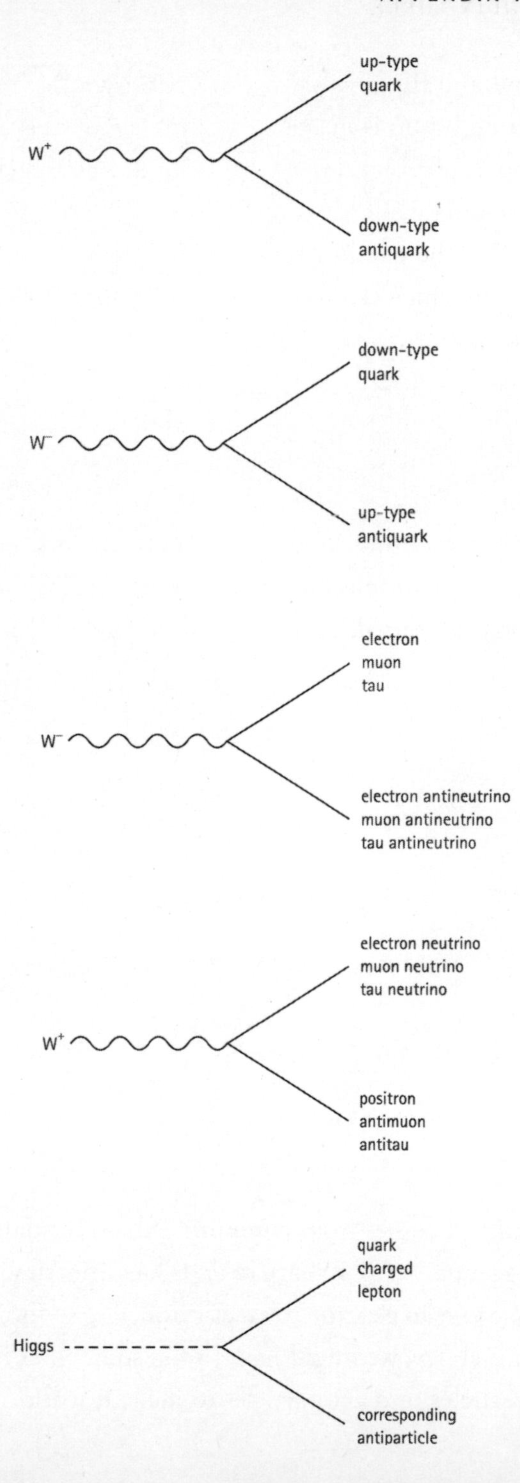
up-type
quark
W⁺
down-type
antiquark
down-type
quark
W⁻
up-type
antiquark
electron
muon
tau
W⁻
electron antineutrino
muon antineutrino
tau antineutrino
electron neutrino
muon neutrino
tau neutrino
W⁺
positron
antimuon
antitau
quark
charged
lepton
Higgs
corresponding
antiparticle

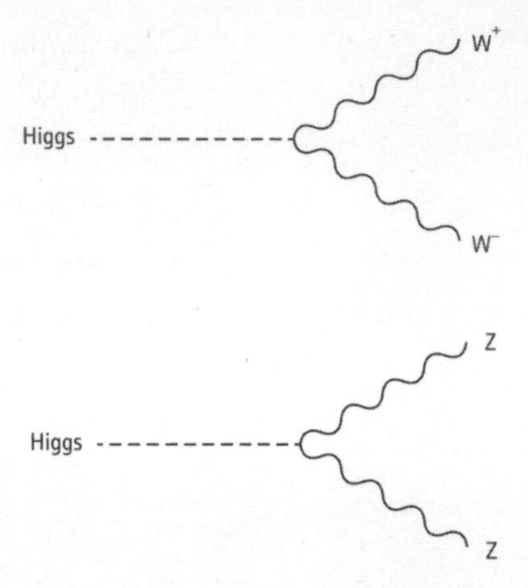

The only remaining fundamental diagram is the Higgs interacting with itself; it can split into two or three copies. Clearly this would violate energy conservation unless it were embedded in a bigger diagram.

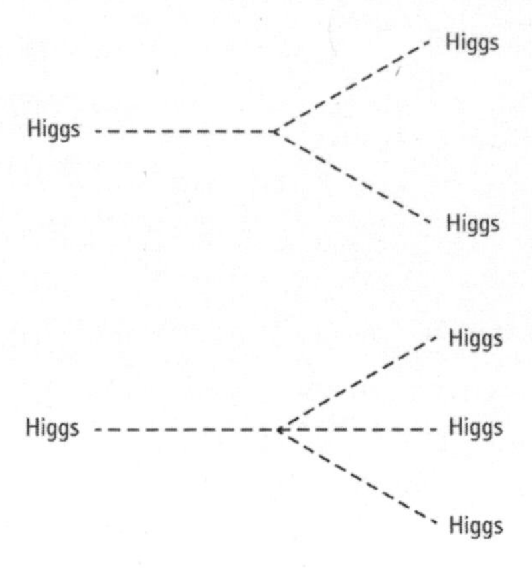

The real fun comes from combining these fundamental diagrams to make bigger ones. All we have to do is join lines describing matching particles: We join an electron to an electron, and so forth. Starting from the diagrams above, we might have to flip some lines from right to left and turn particles into antiparticles to make it work.

For example, let's say we want to ask how a muon can decay. We see that there is a diagram where a muon emits a W^- and turns into a muon neutrino; but that can't happen by itself, since the W is heavier than the muon. Never fear; all is okay as long as the W remains virtual, and decays into something lighter than the muon, such as an electron and its neutrino. All we have to do is glue together the W^- lines from two of the previous diagrams in a consistent way.

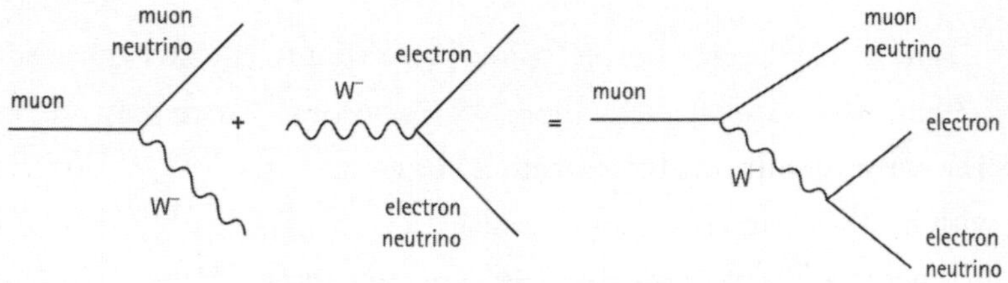

We can also bend lines back on themselves to form loops. Here is a diagram that contributes in an important way to the search for the Higgs at the LHC: a Higgs decaying into two photons. The loop of virtual particles in the middle could contain any particle that couples both to the Higgs (so that the vertex on the left exists) and to photons (so that the vertices on the right exist). Particles with stronger couplings will contribute the most; in this case, that would be the top quark, which is the most massive particle in the Standard Model, and therefore the one with the strongest coupling to the Higgs.

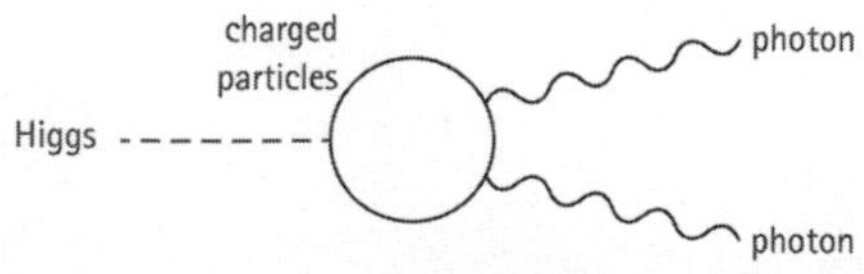

Finally, here are some of the important ways that Higgs bosons are actually produced at the LHC before they decay. There is "gluon fusion," where two gluons come together to make a Higgs; because gluons are massless, they must proceed through a virtual massive particle that feels the strong force, namely a quark.

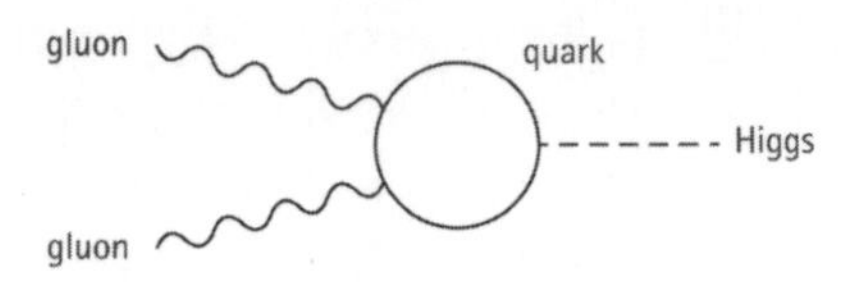

There is also "vector boson fusion," referring to the fact that the W and Z bosons are sometimes called "vector bosons." Since they are massive, they can combine directly into a Higgs.

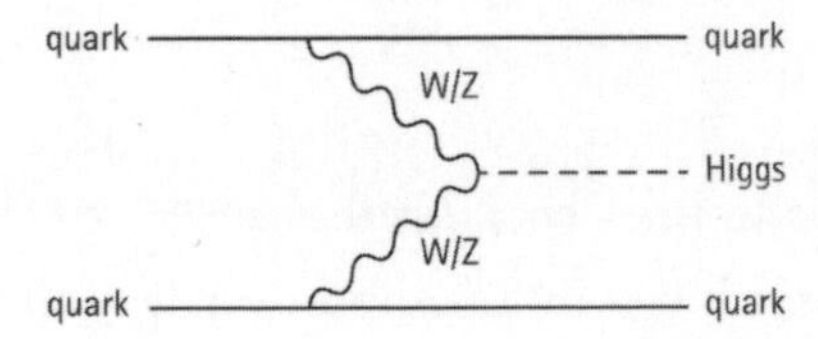

At last there are two different kinds of "associated production," where the Higgs is made along with something else: either a W or Z boson, or a quark-antiquark pair.

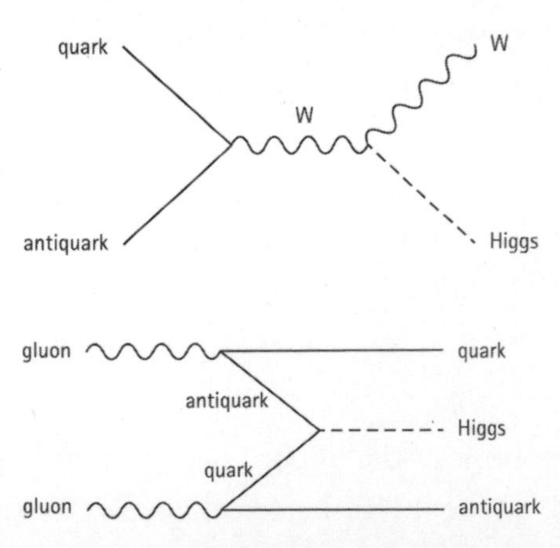

The take-home lesson here isn't the ins and outs of all the different processes that contribute to Higgs production and decay; it's simply that both processes are complicated, arising from a collection of different possibilities, and we have definite rules that allow us to figure out what they are. It's amazing to think that these little cartoons capture something deeply true about the microscopic behavior of the natural world.

FURTHER READING

Aczel, Amir. *Present at the Creation: The Story of CERN and the Large Hadron Collider*. New York: Crown Publishers, 2010.

CERN. CERN faq: LHC, the guide. http://multimedia-gallery.web.cern.ch/multimedia-gallery/Brochures.aspx, 2009.

Close, Frank. *The Infinity Puzzle: Quantum Field Theory and the Hunt for an Orderly Universe*. New York: Basic Books, 2011.

Crease, Robert P., and Mann, Charles C. *The Second Creation: Makers of the Revolution in Twentieth-Century Physics*. New York: Collier Books, 1986.

Halpern, Paul. *Collider: The Search for the World's Smallest Particles*. Hoboken, NJ: Wiley, 2009.

Kane, Gordon. *The Particle Garden: The Universe as Understood by Particle Physicists*. New York: Perseus Books, 1995.

Lederman, Leon, with Teresi, Dick. *The God Particle: If the Universe Is the Answer, What's the Question?* Boston, MA: Houghton Mifflin, 2006.

Lincoln, Don. *The Quantum Frontier: The Large Hadron Collider*. Baltimore, MD: Johns Hopkins University Press, 2009.

Panek, Richard. *The 4 Percent Universe: Dark Matter, Dark Energy, and the Race to Discover the Rest of Reality*. Boston, MA: Mariner Books, 2011.

Randall, Lisa. *Knocking on Heaven's Door: How Physics and Scientific Thinking Illuminate the Universe and the Modern World*. New York: Ecco, 2011.

Sample, Ian. *Massive: The Missing Particle That Sparked the Greatest Hunt in Science*. New York: Basic Books, 2010.

Taubes, Gary. *Nobel Dreams: Power, Deceit, and the Ultimate Experiment.* New York: Random House, 1986.

Traweek, Sharon. *Beamtimes and Lifetimes: The World of High Energy Physicists.* Cambridge, MA: Harvard University Press, 1988.

Weinberg, Steven. *Dreams of a Final Theory.* New York: Vintage, 1992.

Wilczek, Frank. *The Lightness of Being: Mass, Ether, and the Unification of Forces.* New York: Basic Books, 2008.

REFERENCES

References refer to keywords in the main text. The one exception is Chapter Eleven, "Nobel Dreams," where I include two lists of additional references: one for the personal reminiscences of the people involved in the 1964 symmetry-breaking papers, and one that includes all of the technical papers alluded to in the discussion.

Prologue

Hewett: http://blogs.discovermagazine.com/cosmicvariance/2008/09/11/giddy-physicists/

Evans: interview, July 4, 2012.

Higgs: http://www.newscientist.com/article/dn22033-peter-higgs-boson-discovery-like-being-hit-by-a-wave.html?full=true

Chapter One: The Point

Faraday: http://bit.ly/ynX3dL

Heuer: http://www.guardian.co.uk/science/2011/dec/13/higgs-boson-seminar-god-particle

Chapter Two: Next to Godliness

Lederman and Teresi: *The God Particle*, p. xi.

Higgs: http://physicsworld.com/cws/article/indepth/2012/jun/28/peter-higgs-in-the-spotlight

Chapter Four: The Accelerator Story

Janot: V. Jamieson, "CERN Extends Search for Higgs," *Physics World*, October 2000.

Watts: private email, April 4, 2012.

Hewett: interview, February 23, 2012.

Schwitters, Bloembergen: quoted in Kelves, preface to the 1995 edition of *The Physicists: The History of a Scientific Community in Modern America*.

Park: quoted in Weinberg, *Dreams of a Final Theory*, p. 54.

Anderson: Letter to the Editor, *The New York Times*, May 21, 1987.

Krumhansl: Sample, *Massive*, p. 115.

Chapter Five: The Largest Machine Ever Built

Evans, "carnage": interview, July 4, 2012.

Baguette: http://www.telegraph.co.uk/science/large-hadron-collider/6514155/Large-Hadron-Collider-broken-by-bread-dropped-by-passing-bird.html

Evans: http://www.elements-science.co.uk/2011/11/the-man-who-built-the-lhc/

Evans: http://www.nature.com/news/2008/081217/pdf/456862a.pdf

Giudice: *A Zeptospace Odyssey*, pp. 103–104.

Evans, summer party: interview, July 4, 2012.

Chapter Six: Wisdom Through Smashing

Anderson: Eugene Cowan, "The Picture That Was Not Reversed," *Engineering and Science* **46**, 6 (1982).

CERN press release: http://press.web.cern.ch/press/PressReleases/Releases2008/PR10.08E.html

Computing tiers: Brumfield, http://www.nature.com/news/2011/110119/full/469282a.html

Gianotti: interview, May 3, 2012.

Greek Security Team: Roger Highfield, http://www.telegraph.co.uk/science/large-hadron-collider/3351697/Hackers-infiltrate-Large-Hadron-Collider-systems-and-mock-IT-security.html

Chapter Eight: Through a Broken Mirror

Yang and Pauli: Close, *The Infinity Puzzle*, p. 88.

Chapter Nine: Bringing Down the House

Telegraph: http://www.telegraph.co.uk/science/large-hadron-collider/8928575/Search-for-God-Particle-is-nearly-over-as-CERN-prepares-to-announce-findings.html

viXra log: http://blog.vixra.org/2011/12/01/seminar-watch-higgs-special/

CERN update: http://indico.cern.ch/conferenceDisplay.py?confId=150980

Gianotti: http://www.youtube.com/watch?v=0KOoumH4dYA

Gianotti, "Spirit" and "Bear" quotes: interview, May 15, 2012.

Wu: http://physicsworld.com/cws/article/news/2011/dec/14/physicists-weigh-up-higgs-signals

Ellis, Gaillard, and Nanopoulos: *Nuclear Physics B* **106,** 292 (1976).

Britton: http://www.wired.co.uk/news/archive/2011-09/07/david-britton

ATLAS figure: http://www.atlas.ch/news/2012/latest-results-from-higgs-search.html

CMS figure: http://hep.phys.sfu.ca/HiggsObservation/index.php

Megatek: Taubes, *Nobel Dreams*, pp. 137–138.

Higgs: http://www.newscientist.com/article/dn22033-peter-higgs-boson-discovery-like-being-hit-by-a-wave.html?full=true

Incandela: interview, July 4, 2012.

Chapter Ten: Spreading the Word

The Daily Show: http://www.thedailyshow.com/watch/thu-april-30-2009/large-hadron-collider

The Daily Mail: http://www.dailymail.co.uk/sciencetech/article-1052354/Are-going-die-Wednesday.html

Appeals court: http://cosmiclog.msnbc.msn.com/_news/2010/08/31/5014771-collider-court-case-finally-closed?lite

Dorigo: http://www.science20.com/quantum_diaries_survivor/where_will_we_hear_about_higgs_first

Conway 1: http://blogs.discovermagazine.com/cosmicvariance/2007/01/26/bump-hunting-part-1/

Conway 2: http://blogs.discovermagazine.com/cosmicvariance/2007/01/26/bump-huning-part-2/

Conway 3: http://blogs.discovermagazine.com/cosmicvariance/2007/03/09/bump-hunting-part-3/

Cirelli and Strumia: http://arxiv.org/abs/0808.3867

Picozza, Cirelli: http://www.nature.com/news/2008/080902/full/455007a.html
Lykken: http://www.nytimes.com/2007/07/24/science/24ferm.html?pagewanted=all
Woit: http://www.math.columbia.edu/~woit/wordpress/?p=3632&cpage=1#comment-88817
Wu: email, May 2012.
Gianotti: http://www.nytimes.com/2012/06/20/science/new-data-on-higgs-boson-is-shrouded-in-secrecy-at-cern.html?_r=1&pagewanted=all
Schmitt: http://muon.wordpress.com/2012/06/17/do-you-like-to-spread-rumors/
Ouellette: http://news.discovery.com/space/rumor-has-it-120620.html
"Large Hadron Rap": http://www.youtube.com/watch?v=j50ZssEojtM
Kaplan: interview, May 20, 2012.
Particle Fever: http://www.particlefever.com/index.html

Chapter Eleven: Nobel Dreams

Freund: *A Passion for Discovery*, World Scientific (2007).
Anderson: P. W. Anderson, "More Is Different," *Science* **177**, 393 (1972).
Anderson's biggest contribution: email, 2012.
Higgs on Anderson: P. Rodgers, "Peter Higgs: The Man Behind the Boson," *Physics World* **17**, 10 (2004).
Lederman: *The God Particle*.
Lykken: *Symmetry*, http://www.symmetrymagazine.org/cms/?pid=1000087
Bernardi: *Nature*, http://www.nature.com/news/2010/100804/full/news.2010.390.html
Anderson on history: email, 2012.

Personal reminiscences

P. W. Higgs, "Prehistory of the Higgs boson," *Comptes Rendus Physique* **8**, 970 (2007).
P. W. Higgs, "My Life as a Boson," http://www.kcl.ac.uk/nms/depts/physics/news/events/MyLifeasaBoson.pdf (2010).
G. S. Guralnik, "The History of the Guralnik, Hagen, and Kibble Development of the Theory of Spontaneous Symmetry Breaking and Gauge Particles," *International Journal of Modern Physics* **A24**, 2601, arXiv:0907.3466 (2009).

T. W. B. Kibble, The Englert-Brout-Higgs-Guralnik-Hagen-Kibble Mechanism (history)," *Scholarpedia*, http://www.scholarpedia.org/article/Englert-Brout-Higgs-Guralnik-Hagen-Kibble_mechanism_(history)

R. Brout and F. Englert, "Spontaneous Symmetry Breaking in Gauge Theories: a Historical Survey," arXiv:hep-th/9802142 (1998).

Technical articles

V. L. Ginzburg and L. D. Landau, "On the theory of superconductivity," *Journal of Experimental and Theoretical Physics* (USSR) **20**, 1064 (1950).

P. W. Anderson, "An Approximate Quantum Theory of the Antiferromagnetic Ground State," *Physical Review* **86**, 694 (1952).

C. N. Yang and R. L. Mills, "Conservation of Isotopic Spin and Isotopic Gauge Invariance," *Physical Review* **96**, 191 (1954).

L. N. Cooper, "Bound Electron Pairs in a Degenerate Fermi Gas," *Physical Review* **104**, 1189 (1956).

J. Bardeen, L. N. Cooper, and J. R. Schrieffer, "Microscopic Theory of Superconductivity," *Physical Review* **106**, 162 (1957).

J. Bardeen, L. N. Cooper, and J. R. Schrieffer, "Theory of Superconductivity," *Physical Review* **108**, 1175 (1957).

J. Schwinger, "A Theory of the Fundamental Interactions," *Annals of Physics* **2**, 407 (1957).

N. N. Bogoliubov, "A New Method in the Theory of Superconductivity," *Journal of Experimental and Theoretical Physics* (USSR) **34**, 58 [*Soviet Physics-JETP* 7, 41] (1958).

P. W. Anderson, "Coherent Excited States in the Theory of Superconductivity: Gauge Invariance and the Meissner Effect," *Physical Review* **110**, 827 (1958).

P. W. Anderson, "Random-Phase Approximation in the Theory of Superconductivity," *Physical Review* **112**, 1900 (1958).

Y. Nambu, "Quasiparticles and Gauge Invariance in the Theory of Superconductivity," *Physical Review* **117**, 648 (1960).

Y. Nambu and G. Jona-Lasinio, "Dynamical Model of Elementary Particles Based on an Analogy with Superconductivity, I," *Physical Review* **124**, 246 (1961).

Y. Nambu and G. Jona-Lasinio, "Dynamical Model of Elementary Particles Based on an Analogy with Superconductivity, II," *Physical Review* **122**, 345 (1961).

S. L. Glashow, "Partial Symmetries of the Weak Interactions," *Nuclear Physics* **22**, 579 (1961).

J. Goldstone, "Field Theories with Superconductor Solutions," *Nuovo Cimento* **19**, 154 (1961).

J. Goldstone, A. Salam, and S. Weinberg, "Broken Symmetries," *Physical Review* **127**, 965 (1962).

J. Schwinger, "Gauge Invariance and Mass," *Physical Review* **125**, 397 (1962).

P. W. Anderson, "Plasmons, Gauge Invariance, and Mass," *Physical Review* **130**, 439 (1963).

A. Klein and B. Lee, "Does Spontaneous Breakdown of Symmetry Imply Zero-Mass Particles?" *Physical Review Letters* **12**, 266 (1964).

W. Gilbert, "Broken Symmetries and Massless Particles," *Physical Review Letters* **12**, 713 (1964).

F. Englert and R. Brout, "Broken Symmetry and the Mass of Gauge Vector Mesons," *Physical Review Letters* **13**, 321 (1964).

P. W. Higgs, "Broken Symmetries, Massless Particles, and Gauge Fields," *Physics Letters* **12**, 134 (1964).

P. W. Higgs, "Broken Symmetries and the Masses of Gauge Bosons," *Physical Review Letters* **13**, 508 (1964).

A. Salam and J. C. Ward, "Electromagnetic and Weak Interactions," *Physics Letters* **13**, 168 (1964).

G. S. Guralnik, C. R. Hagen, and T. W. B. Kibble, "Global Conservation Laws and Massless Particles," *Physical Review Letters* **13**, 585 (1964).

P. W. Higgs, "Spontaneous Symmetry Breakdown Without Massless Bosons," *Physical Review* **145**, 1156 (1966).

A. Migdal and A. Polyakov, "Spontaneous Breakdown of Strong Interaction Symmetry and the Absence of Massless Particles," *Journal of Experimental and Theoretical Physics* (USSR) **51**, 135 [*Soviet Physics-JETP* **24**, 91] (1966).

T. W. B. Kibble, "Symmetry Breaking in Non-Abelian Gauge Theories," *Physical Review* **155**, 1554 (1967).

S. Weinberg, "A Model of Leptons," *Physical Review Letters* **19**, 1264 (1967).

A. Salam, "Weak and Electromagnetic Interactions," *Elementary Particle Theory: Proceedings of the Nobel Symposium held in 1968 at Lerum, Sweden*, N. Svartholm, ed., p. 367. Almqvist and Wiksell (1968).

G. 't Hooft, "Renormalizable Lagrangians for Massive Yang-Mills Fields," *Nuclear Physics B* **44**, 189 (1971).

G. 't Hooft and M. Veltman, "Regularization and Renormalization of Gauge Fields," *Nuclear Physics B* **44**, 189 (1972).

Chapter Twelve: Beyond This Horizon

Rubin: Ken Croswell. *The Universe at Midnight: Observations Illuminating the Cosmos*. New York: Free Press (2001).

Patt and Wilczek: B. Patt and F. Wilczek, "Higgs-field Portal into Hidden Sectors," http://arxiv.org/abs/hep-ph/0605188

dark-matter collisions with the human body: K. Freese and C. Savage, "Dark Matter Collisions with the Human Body," http://arxiv.org/abs/arXiv:1204.1339

"Higgs in Space": C. B. Jackson, et al., "Higgs in Space," *Journal of Cosmology and Astroparticle Physics* **4**, 4 (2010).

Shaposhnikov and Tkachev: M. Shaposhnikov and I.I. Tkachev, "Higgs Boson Mass and the Anthropic Principle," *Modern Physics Letters* A **5,** 1659 (1990).

106 GeV: B. Feldstein, L. Hall, and T. Watari, "Landscape Predictions for Higgs Boson and Top Quark Masses," *Physical Review* D **74,** 095011 (2006).

Weinberg: S. Weinberg, *Physical Review Letters* **59**, 2607 (1987).

Chapter Thirteen: Making It Worth Defending

Wilson: http://blogs.scientificamerican.com/cocktail-party-physics/2011/09/23/protons-and-pistols-remembering-robert-wilson/

Weinberg: http://www.nybooks.com/articles/archives/2012/may/10/crisis-big-science/

National Journal: http://news.nationalpost.com/2012/07/05/higgs-boson-find-could-make-light-speed-travel-possible-scientists-hope/

Mansfield 1: E. Mansfield, "Academic Research and Industrial Innovation," *Research Policy* **20**, 1 (1991).

Mansfield 2: E. Mansfield, "Academic Research and Industrial Innovation: An Update of Empirical Findings," *Research Policy* **26**, 773 (1998).

Cartoon: Z. Weiner, *Saturday Morning Breakfast Cereal*, http://www.smbc-comics.com/index.php?db=comics&id=2088

Yahia: http://blogs.nature.com/houseofwisdom/2012/07/the-social-aspect-of-the-higgs-boson.html

Evans: interview, July 4, 2012.

Appendices

For more on helicity, see F. Tanedo, "Helicity, Chirality, Mass, and the Higgs," http://www.quantumdiaries.org/2011/06/19/helicity-chirality-mass-and-the-higgs/

ACKNOWLEDGMENTS

I make my living as a physicist, but my specialty is theoretical gravitation and cosmology; in particle physics I am a semi-tourist, and I haven't been involved directly in an experiment since I was an undergraduate. I owe an enormous debt to a large number of people who generously helped me during this project, both by sharing their insights and by reading drafts of the book.

A number of physicists who work on this stuff for a living were kind enough to be interviewed for this book, either by phone or by email. It's a pleasure to thank Philip Anderson, John Conway, Gerald Guralnik, Fabiola Gianotti, JoAnne Hewett, Joe Incandela, Gordy Kane, David Kaplan, Mike Lamont, Joe Lykken, Jack Steinberger, Gordon Watts, Frank Wilczek, and Sau Lan Wu for enormously helpful conversations. Mistakes are all completely my fault, needless to say—and my apologies for using only a tiny fraction of the stories I was told.

I was also fortunate enough to get help from both professional physicists and amateur lovers of science who answered specific questions or offered comments on the text. Big thanks to Allyson Beatrice, Dan Birman, Matt Buckley, Alicia Chang, Lauren Gunderson, Kevin Hand, Ann Kottner, Rick Loverd, Rusi Mchedlishvili, Philip Phillips, Abbas Raza, Henry Reich, Ira Rothstein, Maria Spiropulu, David

Saltzberg, Matt Strassler, and Zach Weinersmith for spending time reading the book and offering input. Their comments have improved the manuscript a millionfold. Special thanks to Zach for sharing the comic reprinted in the insert, which says it all.

Thanks to my students and collaborators, who once again showed great patience with me when I would disappear for lengthy stretches of time. (At least they seemed patient from where I was sitting.) And let me send my appreciation to all the readers of our blog, *Cosmic Variance*, and everyone who comes to hear me talk about these topics in public lectures. I am constantly amazed and delighted at the genuine enthusiasm for science and learning that I encounter on a regular basis.

Without my editor, Stephen Morrow, and the good folks at Dutton, this book would have likely never been instigated, and if it had it wouldn't have been nearly as good. Without my agents, Katinka Matson and John Brockman, I probably wouldn't be writing books in the first place.

In the dedication to their famous textbook *Gravitation*, Charles Misner, Kip Thorne, and John Wheeler express their thanks to their fellow citizens for supporting public expenditures on science. For giant projects like the Large Hadron Collider, more than a little bit of government spending is required, as well as an impressive amount of international collaboration. Sincere thanks to all the people of all the countries of the world who help enable the quest to discover nature's deepest secrets. Reporting back on the wonders we have found is really the least we can do.

I fell in love with the talented writer Jennifer Ouellette because of her good looks, piercing intellect, and engaging personality, not because she is endlessly patient and extremely helpful when it comes to writing books. But it is a nice side benefit. My eternal love and appreciation.

INDEX

Note: page numbers in *italics* indicate charts and illustrations.